Nanostructure Science and Technology

Series editor
David J. Lockwood, FRSC
National Research Council of Canada
Ottawa, Ontario, Canada

More information about this series at http://www.springer.com/series/6331

Huisheng Peng

Fiber-Shaped Energy Harvesting and Storage Devices

Springer

Huisheng Peng
Department of Macromolecular Science
Fudan University
Shanghai, China

ISSN 1571-5744 ISSN 2197-7976 (electronic)
Nanostructure Science and Technology
ISBN 978-3-662-45743-6 ISBN 978-3-662-45744-3 (eBook)
DOI 10.1007/978-3-662-45744-3

Library of Congress Control Number: 2014960207

Springer Heidelberg New York Dordrecht London

Printed on acid-free paper

Springer-Verlag GmbH Berlin Heidelberg is part of Springer Science+Business Media (www.springer.com)

Preface

The effective energy harvest and storage is critical to the modern society. For the energy harvest, solar energy that is inexhaustible, free, environmentally friendly and uniformly distributed to all countries represents one of the most explored strategies, and a solar cell that converts solar energy to electrical energy has been widely investigated. The silicon-based solar cells are already available for practical applications, but the high cost and pollution prevent their large spread. As a result, a lot of efforts have been made to develop next-generation solar cells including dye-sensitized, polymer and perovskite solar cells in recent decades. They are expected to dominate the future life due to the environmental friendliness and low cost. For the energy storage, electrochemical storage devices such as lithium ion batteries and supercapacitors are widely used to power various electronic facilities. Currently, main attentions are paid to discover new materials and design new structures to further improve the energy storage capability.

Generally, the solar cells, lithium ion batteries and supercapacitors appear in a rigid plate or flexible film that cannot effectively meet the combined requirement on the light weight, miniaturization and weaveability in the modern electronics such as wearable facilities, the proposed next-generation resolution in the near future. The large and heavy lithium ion battery at the leg end of Google glasses makes them inconvenient and uncomfortable. It even becomes a bottleneck for wearable electronic products such as smart clothes. To this end, a new family of fiber-shaped energy devices has been recently discovered as an effective solution. Compared with the conventional two- or three-dimensional structures, the one-dimensional energy devices show some unique and promising advantages including smaller size, lighter weight, three-dimensional flexibility and weaveability.

This book has summarized the advancement of fiber-shaped energy devices mainly based on the research work at our lab. To realize these one-dimensional energy devices with high performances, it is critically important to develop desirable fiber electrodes. Therefore, the available fiber electrodes are first compared and a new family of aligned carbon nanotube-based fibers are recommended for both energy conversion and storage devices. Based on the aligned carbon nanotube fiber as one or two electrodes, fiber-shaped dye-sensitized solar cells, polymer solar cells and perovskite solar cells are sequentially investigated. Two main structures, i.e., twisted and coaxial, have been described for the three kinds of solar cells with an emphasis on their advantages and disadvantages. Fiber-shaped energy storage devices including lithium ion batteries and supercapacitors are then followed on the basis of similar fiber electrodes and structures. To realize self-powering functionality that is highly desired in the wearable facilities as well as other portable electronic products, fiber-shaped solar cells are further integrated with fiber-shaped energy storage devices. In other words, both photoelectric conversion and electrochemical storage are achieved at a single fiber device. Finally, the above fiber-shaped energy devices have been woven into flexible energy textiles, and the future directions on these wearable energy devices are also highlighted.

This book is intended for scientists, engineers, graduate students and undergraduate students, majoring in chemistry, physics, biomedical science and engineering who are interested in the energy materials and devices. It may be also useful to many non-specialists in industry who are devoted to promoting the wearable and related technologies. As fiber-shaped energy devices have been studied just for a few years, there may be many imperfections and omissions in this book also due to the limit of my ability. I will greatly appreciate critical and suggestive comments from the readers.

I strongly hope to acknowledge many postdoctoral fellows and graduate students who joined in the writing of this book at my lab with Chapter 1 drafted by Hao Sun, Chapter 2 by Longbin Qiu, Chapter 3 by Shaowu Pan and Xin Fang, Chapter 4 by Zhitao Zhang, Chapter 5 by Jue Deng and Longbin Qiu, Chapter 6 by Houpu Li, Huijuan Lin and Guozhen Guan, Chapter 7 by Wei Weng and Ye Zhang, Chapter 8 by Hao Sun and Jue Deng, Chapter 9 by Shaowu Pan and Chapter 10 by Hao Sun. In particular, I would like to thank Hao Sun for the organization work during writing and Xin Fang for revising the whole book at the final stage. Yifan Xu, a graduate student in my lab, also contributed to the book by drafting several illustrations. I sincerely thank Editor June Tang for the kind invitation on writing this book and warm support in publishing the book.

Shanghai, China
2014

Huisheng Peng

Contents

Chapter 1
Introduction

Abstract In this chapter, the necessity for the fiber-shaped device is discussed in modern electronics. The main efforts are first paid to describe the difficulty in the development of conventional planar energy harvesting and storage devices including solar cells, electrochemical capacitors, and lithium ion batteries. The history and recent advancement in the fiber-shaped device are then summarized along three directions in energy conversion, energy storage, and device integration.

1.1 Development in Energy Harvesting and Storage

1.1.1 Energy Harvesting

Energy crisis has forced people to find renewable and clean energy resources. The commercial silicon-based solar cells show a high energy conversion efficiency while suffer from a complex fabrication process and high cost. In addition, they are generally made into rigid devices which are unfavorable for the flexible and portable electronics. As a new-generation solar cell, dye-sensitized solar cell has attracted the interests from both academics and industries since they were presented by M. Grätzel and B. O'Regan in 1991, due to the advantages including easy fabrication, low cost, high performance in lower light intensity, and the potential to be flexible [1]. The highest power conversion efficiency of dye-sensitized solar cell has achieved 13 % in the past two decades, which are very competitive with conventional silicon-based solar cells.

Polymer solar cells are also attractive based on a number of advantageous features, including ultrathin-film architecture, low material consumption resulting from a high absorption coefficient, abundant material resource, easy fabrication based on a solution method, and flexibility. Materials with a delocalized π electron system can absorb sunlight to create and transport photogenerated charge carriers. Research on polymer solar cell generally focuses on solution processable organic semiconducting polymers. Since the photoinduced phenomenon of conjugated polymer and C_{60} was reported by N. S. Sariciftci in 1992, the efficiency of the polymer photovoltaic device has been improved significantly [2]. Currently, a maximum efficiency of 10.6 % has been achieved by designing a new low bandgap polymer [3].

H. Peng, *Fiber-Shaped Energy Harvesting and Storage Devices*, Nanostructure Science and Technology, DOI 10.1007/978-3-662-45744-3_1

1.1.2 *Energy Storage*

Energy storage is another severe challenge in the twenty-first century. The modern society and the increasing emerging ecological problem badly require low-cost and environmentally friendly energy storage devices with high performance. The prosperity of portable electronics and electric vehicles market stimulates the development of energy storage devices, towards higher power density and energy density. Electrochemical capacitors and batteries are two typical energy storage devices, which are comprised of two electrodes, a separator and electrolyte. The electrical energy is stored in electrochemical processes. Specifically, during charge and discharge process, the charged ions shuttle between two electrodes, generating current through the external circuit. The capability of storing energy strongly relies on the properties of electrodes, including micro-architecture, conductivity, specific surface area, and electrochemical performances.

Electrochemical capacitor, also known as supercapacitor, is considered as one of the most efficient energy storage devices. The conception was first proposed by Becker in 1957 in a patent filed and then commercialized by Panasonic in 1978 [4]. Compared with lithium ion batteries, supercapacitors represent the following advantages: (1) an outstanding power-delivery capability (high power density), (2) a fast-charging process (short time needed for full charging), (3) a long cyclic stability (charging–discharging for thousand times without obvious decay), and (4) a high coulombic efficiency. Although most electrochemical capacitors show a lower energy density, they can be integrated with lithium ion batteries in practical applications, e.g., for hybrid electric vehicles, electrochemical capacitors can provide instantaneous power for acceleration or climbing, and they will be recharged during regenerative braking.

Lithium ion battery is another energy storage device that is widely used. Compared with traditional secondary batteries such as lead–acid or nickel–cadmium batteries, rechargeable lithium ion batteries exhibit superior performance including long cycle life, high specific energy density, and no memory effect. The metallic lithium can be replaced by carbon matrixes that can reversibly host and release lithium ions at low electrochemical potentials. In this case, lithium ions will shuttle between cathode (e.g., $LiCoO_2$) and anode (e.g., graphite) realizing energy storage and release during charge and discharge process.

1.2 Challenges for Conventional Planar Devices

Although various achievements have been made in conventional planar devices for energy harvesting and storage, the development of planar devices still faces some challenges. As for dye-sensitized solar cells and polymer solar cells, a promising advantage over the silicon solar cell lies in that they can be fabricated into a flexible configuration on plastic conductive substrates. However, the plastic conductive substrates cannot sustain high sintering temperature for semiconducting metal oxide (e.g., 450–500 °C for TiO_2) during fabrication. The flexible polymer solar cells can be fabricated at a large scale by screen printing or roll-to-roll process, but the energy conversion efficiency is relatively low. In addition, the developing wearable electronics badly needed the supporting energy harvesting and storage devices with high flexibility, lightweight, wearability, and high performance. However, the planar devices can hardly satisfy the above requirements. To this end, to develop efficient fiber-shaped devices which are flexible, lightweight, and weavable is an ideal strategy to solve the mentioned challenges. Since the conception of fiber-shaped dye-sensitized solar cell emerged in 2001, fiber-shaped devices for energy harvesting, energy storage, and device integration have been materialized, and various interesting and encouraging results have been achieved.

1.3 Emergence of Fiber-Shaped Devices

1.3.1 Fiber-Shaped Energy Harvesting Devices

The emergence of fiber-shaped energy harvesting devices can be traced as early as 2001 [5]. The configuration of this novel fiber-shaped dye-sensitized solar cell, or as they called "solar cell in a fiber form," was similar with currently adopted configuration. In this coaxial structure, a stainless steel wire was covered with a layer of dye-absorbed TiO_2 particles as working electrode, and a transparent conductive polymer, as counter electrode, was wrapped outside. Electrolyte was injected into a transparent tube before sealing. The as-fabricated solar cell, 10 cm in length, was able to deliver a voltage of 0.3–0.35 V. Although the performance was limited by the poor conductivity of the conducive polymer layer and corrosive electrolyte, the first fiber-shaped dye-sensitized solar cell, as well as the first fiber-shaped device, has shed light on the possibility in fabricating wearable electronics and inspired more scientists in the energy field to develop fiber-shaped devices with high performances, as well as preparing superior wire electrodes.

As another important new-generation solar cell, the first fiber-shaped polymer solar cell came up later in 2007 [6]. Optical fibers have been used as substrates in the core sheath architecture as the incident light can be conducted into the photoactive layer along the coaxial direction. The conductive indium tin oxide layer and photoactive layer of P3HT or PCBM were sequentially coated on the optical fiber by a dip-coating process, and the aluminum layer was then deposited on the outer surface by thermal evaporation method. The reflective and transmission losses can be minimized, and an energy conversion efficiency of 1.1 % was obtained. Since then, various achievements have been made and will be introduced in Chap. 4.

1.3.2 Fiber-Shaped Energy Storage Devices

In 2002, the first fiber-shaped energy storage device was fabricated by Baughman and coworkers using single-walled carbon nanotube fibers as electrodes. The resulting supercapacitor exhibited a specific capacitance of 5 F g^{-1} and an energy storage density of 0.6 Wh kg^{-1}, which, moreover, can be well maintained over 1,200 cycles [7]. The fiber-shaped supercapacitors were further woven into textiles as a paradigm. This pioneering work motivated the investigation on fiber-shaped energy storage, and the carbon nanotube fiber was intensely explored as a promising electrode material based on its high flexibility, strength, and conductivity.

Fiber-shaped supercapacitor shares the advantages of high power density (10 kW kg^{-1}), fast charging and discharging, and long cycle life ($>10^5$ cycles). However, its relatively low energy density has hindered their applications. To this end, developing fiber-shaped lithium ion battery is very useful to power wearable electronics. In 2005, Gordon G. Wallace et al. fabricated the first fiber-shaped lithium ion polymer fiber battery [8]. Both electrodes used the platinum wire as the substrate where polypyrrole–hexafluorophosphate (cathode) and polypyrrole–polystyrenesulfonate (anode) are in situ electropolymerized. Noted that the platinum wire was previously wounded onto a hollow poly(vinylidene) fluoride (PVDF) membrane scroll. The cathode was inserted into the hollow PVDF scroll. Here, the PVDF was used as the separator to prevent the contact of two electrodes. The battery was sealed into a glass vial filled with electrolyte. The discharge capacity of the whole battery is about 10 mAh g^{-1} and can be maintained over 30 cycles. Besides, the discharge voltage plateau is approximately 1 V and the coulombic efficiency is about 98 %. They also examined the flexibility of this battery. After tied into a knot, the battery showed a similar capacity as before, which indicates its good flexibility. This work presented the first attempt towards high-performance fiber-shaped lithium ion battery. Up to date, metal electrode-based and aligned carbon nanotube fiber-based lithium ion batteries have been successfully fabricated with large discharge capacity, which will be discussed in detail in Chap. 7.

1.3.3 Fiber-Shaped Integrated Devices

Fiber-shaped energy harvesting and storage devices have been independently investigated for a couple of years. The requirement of storing the harvested energy in an energy storage device has driven the development of fiber-shaped integrated device. The first fiber-shaped integrated device was reported by Wang and coworkers [9]. It incorporated a dye-sensitized solar cell, a nanogenerator, and a supercapacitor to realize the energy harvest and storage in a single fiber-shaped device. The exciting conception and results have encouraged more efforts to fabricate fiber-shaped integrated devices, and various improvements have been achieved in the structure, stability performance of the integrated device, which will be introduced in Chap. 8.

1.4 Advancement of Fiber-Shaped Device

In the past 10 years, substantial progresses have been achieved in the field of device structure, fabrication technique, electrode material, and device functionalization of the fiber-shaped devices. For fiber-shaped dye-sensitized solar cells, the highest energy conversion efficiency has achieved 8.45 %, and multifunctions including stretchability and magnetic response have been realized to meet the requirements in practical applications [10–12]. For fiber-shaped polymer solar cells, the highest energy conversion efficiency has achieved 3.81 % [13]. The fiber-shaped polymer solar cells enable the fabrication of all-solid-state devices, which are favorable in wearable applications.

The investigations on fiber-shaped supercapacitor mainly focus on new electroactive materials and novel device configurations to increase the specific capacitance of the device. Currently, fiber-shaped supercapacitors in coaxial and twisted configuration have been fabricated with high performance. To further increase the specific capacitance, one of the efficient strategies is to introduce active materials with pseudocapacitance, such as metal oxides (MnO_2, RuO_2), conductive polymers (polyaniline, polypyrrole, and poly(3,4-ethylenedioxythiophene)), and doped carbon materials (N-doped graphene). With the incorporation of these materials, the specific capacitance has been largely increased. For example, a recent report presented a fiber-shaped supercapacitor with a large specific volume capacitance of 300 F cm^{-3} and energy density of 6.3 mWh cm^{-3}. The energy density is comparable to a thin-film lithium battery with an output voltage of 4 V and capacity of 500 mAh, but the power density is two orders of magnitude higher [14]. In addition, some attempts have been made to weave fiber-shaped supercapacitors into flexible textiles, which shed light on the potential use in wearable applications in the future.

The development of fiber-shaped lithium ion battery has just started for a few years [8]. Up to date, the specific mass capacity of fiber-shaped lithium ion battery is comparable to its planar analogues, but the capacity of the entire device is still low, which calls for scaling up the fiber-shaped lithium ion battery and more loading of active materials. A good news is that some attempts have been made to weave stretchable fiber-shaped lithium ion batteries into textiles, realizing stable power supply as stretchable and wearable electronics [15].

The research on fiber-shaped integrated device was also started lately. However, based on the achievements on fiber-shaped solar cells and supercapacitors, rapid progress has been achieved in the past 3 years. Both coaxial and twisted configurations have been realized [16, 17], as well as all-solid-state devices based on fiber-shaped polymer solar cells [18]. Stretchable fiber-shaped integrated device has also been developed based on stretchable fiber-shaped solar cell and supercapacitor with a coaxial structure. We believe there are still plenty of opportunities to further optimize the structure, material, and technique to fabricate better fiber-shaped integrated devices.

References

1. Oregan B, Gratzel M (1991) A low-cost, high-efficiency solar-cell based on dye-sensitized colloidal TiO_2 films. Nature 353(6346):737–740
2. Sariciftci NS, Smilowitz L, Heeger AJ, Wudl F (1992) Photoinduced electron transfer from a conducting polymer to buckminsterfullerene. Science 258(5087):1474–1476
3. You J, Dou L, Yoshimura K, Kato T, Ohya K, Moriarty T, Emery K, Chen C-C, Gao J, Li G, Yang Y (2013) A polymer tandem solar cell with 10.6 % power conversion efficiency. Nat Commun 4:1446
4. Dai LM, Chang DW, Baek JB, Lu W (2012) Carbon nanomaterials for advanced energy conversion and storage. Small 8(8):1130–1166
5. Baps B, Eber-Koyuncu M, Koyuncu M (2001) Ceramic based solar cells in fiber form. Key Eng Mater 206:937–940
6. Liu J, Namboothiry MAG, Carroll DL (2007) Optical geometries for fiber-based organic photovoltaics. Appl Phys Lett 90(13):133515
7. Dalton AB, Collins S, Munoz E, Razal JM, Ebron VH, Ferraris JP, Coleman JN, Kim BG, Baughman RH (2003) Super-tough carbon-nanotube fibres. Nature 423(6941):703–703
8. Wang JZ, Too CO, Wallace GG (2005) A highly flexible polymer fibre battery. J Power Sources 150:223–228
9. Bae J, Park YJ, Lee M, Cha SN, Choi YJ, Lee CS, Kim JM, Wang ZL (2011) Single-fiber-based hybridization of energy converters and storage units using graphene as electrodes. Adv Mater 23(30):3446–3449
10. Yang Z, Sun H, Chen T, Qiu L, Luo Y, Peng H (2013) Photovoltaic wire derived from a graphene composite fiber achieving an 8.45 % energy conversion efficiency. Angew Chem Int Ed 52(29):7545–7548
11. Yang Z, Deng J, Sun X, Li H, Peng H (2014) Stretchable, wearable dye-sensitized solar cells. Adv Mater 26(17):2643–2647
12. Sun H, Yang Z, Chen X, Qiu L, You X, Chen P, Peng H (2013) Photovoltaic wire with high efficiency attached onto and detached from a substrate using a magnetic field. Angew Chem Int Ed 52(32):8276–8280
13. Lee MR, Eckert RD, Forberich K, Dennler G, Brabec CJ, Gaudiana RA (2009) Solar power wires based on organic photovoltaic materials. Science 324(5924):232–235
14. Yu DS, Goh K, Wang H, Wei L, Jiang WC, Zhang Q, Dai LM, Chen Y (2014) Scalable synthesis of hierarchically structured carbon nanotube-graphene fibres for capacitive energy storage. Nat Nanotechnol 9(7):555–562
15. Ren J, Zhang Y, Bai W, Chen X, Zhang Z, Fang X, Weng W, Wang Y, Peng H (2014) Elastic and wearable wire-shaped lithium-ion battery with high electrochemical performance. Angew Chem Int Ed 53(30):7864–7869
16. Chen X, Sun H, Yang Z, Guan G, Zhang Z, Qiu L, Peng H (2014) A novel "energy fiber" by coaxially integrating dye-sensitized solar cell and supercapacitor. J Mater Chem A 2(6):1897
17. Chen T, Qiu L, Yang Z, Cai Z, Ren J, Li H, Lin H, Sun X, Peng H (2012) An integrated "energy wire" for both photoelectric conversion and energy storage. Angew Chem Int Ed 51 (48):11977–11980
18. Zhang Z, Chen X, Chen P, Guan G, Qiu L, Lin H, Yang Z, Bai W, Luo Y, Peng H (2014) Integrated polymer solar cell and electrochemical supercapacitor in a flexible and stable fiber format. Adv Mater 26(3):466–470

Chapter 2
Electrically Conducting Fiber

Abstract The transformation towards fiber shape starts from the fiber electrode. In this chapter, the requirements for fiber electrodes in fiber-shaped devices are first discussed. A wide variety of inorganic and organic materials as well as their composites are then compared for the preparation and property of fiber electrodes. Besides the conventional metals, more efforts are made to discuss various nanomaterial fibers based on nanostructured carbon such as carbon nanotube and graphene and polymers.

2.1 Introduction

Conventional electronic and energy devices are constructed in a planar structure on rigid substrate or flexible polymer film [1, 2]. Many efforts have been devoted to improve the performance and extend functions by designing new materials and integrating functional devices [3, 4]. However, the planar structure is not suitable for arbitrary surface and cannot endure twisting deformation, which limits their applications. For extended requirement, research in flexible and wearable electronics has gained increasing popularity in recent years, which necessitates the development of fiber-shaped devices [5, 6]. As a prosperous research area, fiber-shaped devices are highly desirable for wearable electronics that are expected to be lightweight, flexible, and adaptable. In the recent years, many fiber-shaped devices have been fabricated through well-established and low-cost technologies, normally at ambient conditions. The advanced nanotechnology has made it feasible to assemble nanomaterials or devices directly on a single fiber, which typically has a thickness of several to tens of micrometers. Nowadays, the mainly investigated fiber-shaped energy devices include polymer solar cells, dye-sensitized solar cells, perovskite solar cells, supercapacitors, and lithium ion batteries.

It is incontrovertible that the electrode plays an essential role in the performance of energy devices [7]. High conductivity is in favor of charge collection and transport which is the prerequisite for electrode in energy devices. The electrode materials in conventional structure devices vary from metal, semiconductors, carbon materials, conducting polymers, and their composites, which normally are

H. Peng, *Fiber-Shaped Energy Harvesting and Storage Devices*, Nanostructure Science and Technology, DOI 10.1007/978-3-662-45744-3_2

reluctant to be flexible and deformable. The key issue to fabricate the fiber-shaped device is preparing fiber electrodes which are conductive, electrochemical active, and flexible. In this chapter, we will present a series of electrically conducting fiber made from metal, carbon, and polymers, which lay the foundation of the fiber-shaped energy devices.

Generally, a conducting fiber is not capable of serving as an electrode individually. Active materials are always required for composition. As a critical part in fiber-shaped device, a fiber electrode should possess suitable conductivity, mechanical strength, flexibility, and electrochemical activity. A variety of materials have been prepared into fiber shape and used as fiber electrodes including metal wire, carbon-based fiber, polymer fiber, and composite fiber. Each of them shows merits and faults in different circumstances. For example, metal wires have high conductivity, which is vital for charge transport, diminishing internal resistance and facilitating electrochemical kinetics. But, on another aspect, they suffer from high expenses and heavy weight. Moreover, the stiffness and rigidity of the metal wire will cause interior stress during assembly, thus weakening the flexibility of the device. Polymer fibers serve as flexible and elastic substrates, but their intrinsic nonconductivity blocks their application. Various carbon nanomaterials like carbon nanotube and graphene, exhibiting excellent conductivity, mechanical strength, electrochemical properties, and high specific surface area, become arresting candidates in fiber-shaped devices, which will be highlighted in this chapter.

2.2 Metal Wire

Metal wire has widely been used in cable wire in daily life due to its high electrical conductivity. Table 2.1 exhibits the mechanical and electrical properties of commonly used metal materials. In fiber-shaped devices, titanium wire and stainless steel wire are usually favored as conducting substrates and electrodes.

Table 2.1 Properties of common metal materials

Entries	Density ($g\ cm^{-3}$)	Young modules (GPa)	Conductivity ($10^6\ S\ m^{-1}$)
Titanium (Ti)	4.5	116	2.4
Stainless steel (type 304)	7.8	211	10.4
Platinum (Pt)	21.4	168	9.5
Nickel (Ni)	8.9	200	14.4
Gold (Au)	19.3	79	45.1
Silver (Ag)	10.5	83	63.0
Aluminum (Al)	2.7	70	35.5
Copper (Cu)	8.9	110–128	58.8

2.2.1 Titanium Wire

Titanium (Ti) is a lustrous transition metal with many excellent properties. It has a low density, a high mechanical strength, conductivity, and melting point (1,942 K). Although these merits are shared by other metals, the indispensability of Ti wire in fiber-shaped devices lies in the intrinsic advantage that it benefits TiO_2 modification on the surface.

As a semiconductor, TiO_2 is widely used in photocatalysis, photovoltaic, water splitting, electrochromis devices, and hydrogen storage [8]. TiO_2 nanomaterials, with various architectures like nanoparticles, nanotubes, or nanowires, can be readily deposited or in situ synthesized on Ti wire through many methods including hydrothermal, dip-coating, or anodization process [9–14], which can serve as working electrode in dye-sensitized solar cells, polymer solar cells, perovskite solar cells, and supercapacitors. The properties of TiO_2 rely on its nanostructure morphology and arrangement in assemblies, which is affected by preparing process. As a demonstration, Fig. 2.1b shows a typical scanning electron microscopic (SEM) image of TiO_2 nanoparticles dip coated from TiO_2 slurry. At first glance, the TiO_2 is uniform, but the magnified image shows that TiO_2 coating is porous with

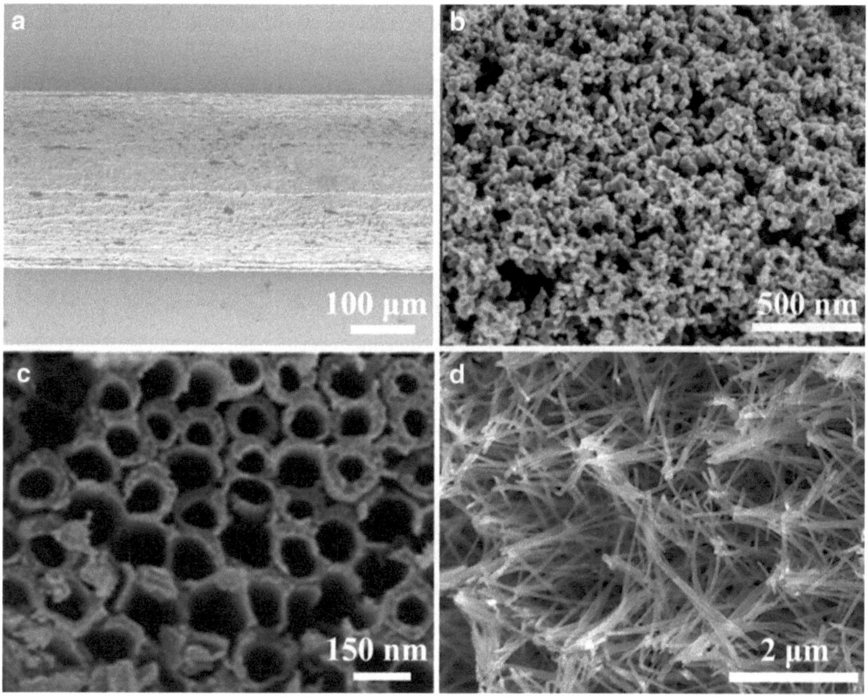

Fig. 2.1 Scanning electron microscopic (SEM) images of TiO_2 nanomaterials coated on a Ti wire. **a** Bare Ti wire. **b** TiO_2 nanoparticles. **c** TiO_2 nanotubes (Reproduced from Ref. [13] by permission of John Wiley & Sons Ltd). **d** TiO_2 nanowire (Reproduced from Ref. [9] by permission of The Royal Society of Chemistry)

nanoparticles aggregated, providing a sufficient surface area. However, the TiO_2 coating layer is brittle and apt to divorce from substrate arising from the weak adhesion, especially under deformation. It has been illustrated that thickness of coating layer has a huge impact [15]. For a thin coating layer, cracks appear at the superficial area, and large patches of coating are still anchored on the Ti wire, while in a thick coating layer, the detachment takes place at the interface closed to the cracks. TiO_2 coating layer provides a high surface area, but it is susceptible to deformation such as bending and stretching, rendering the TiO_2nanoparticle-coated Ti wire incompatible to fiber-shaped device. Apart from the structural instability, TiO_2 nanoparticles are randomly aggregated, which impedes the charge transport and introduces broad energetic disorder [16].

On this account, alternative architectures of TiO_2, with controllable morphology and alignment, such as TiO_2 nanorods and nanotubes, become appealing [8]. In principle, their one-dimensional (1D) configuration provides effective pathways for charge transport and reduces the crystal boundaries that are ubiquitous in TiO_2 nanoparticles. TiO_2 nanotubes can be in situ synthesized on the Ti wire through electrochemical anodization, with morphology tailored by preparing parameters [17]. Figure 2.1c shows SEM images of a TiO_2nanotube-modified Ti wire. The TiO_2 nanotubes are vertically aligned on the Ti surface, making the Ti wire sustainable for deformation. Similarly, TiO_2 nanowires, prepared through an alkali hydrothermal method, also exhibit one-dimensional configuration, which profits for charge transport and reducing the charge recombination (Fig. 2.1d) [18].

Apart from TiO_2, other inorganic oxide, like Co_3O_4, can also be coated around Ti wire using similar methods. For example, porous Co_3O_4 nanowires were grown on a Ti wire through a solvothermal method which can be used in fiber-shaped supercapacitors [19].

2.2.2 *Stainless Steel Wire*

Stainless steel, commercially known as inox steel, is a steel alloy with 10.5 wt% of chromium (Cr). As its name suggests, stainless steel is resistive to corrosion from water. But it is not absolutely stain-proof, especially in saline environments. Stainless steel is differentiated from carbon steel by the chromium content. Unprotected carbon steel is prone to rust when exposed in air and moisture. The Fe_xO_y film (the rust) is active and accelerates corrosion by producing more Fe_xO_y, which is apt to flake and fall away. Stainless steels contain sufficient Cr to form a passive film of Cr_xO_y, which prevents further surface corrosion by blocking oxygen permeation. Stainless steel with excellent conductivity can be used in cable wires as well as fiber electrode in fiber-shaped devices [20]. Active components like TiO_2, ZnO, and Pt particles and silver film can be introduced through dip coating [20–23]. Introducing a second component can modify the work function and provide sufficient surface area and catalytic site for electrochemical applications.

2.2.3 Other Metal Wires

Besides the popular Ti wire and stainless steel wire, other metal wires including nickel (Ni) wire, copper (Cu) wire, and noble metal wires have also been employed in fiber-shaped devices.

Nickel is a transition metal with silvery luster. It is hard, ductile, and corrosion resistant at room temperature. Many materials like $Ni(OH)_2$ and Co_3O_4 nanowires can be incorporated with a nickel wire to perform as fiber electrodes in fiber-shaped energy devices (Fig. 2.2a, b) [19, 24].

Copper is a ductile metal with very high thermal and electrical conductivity. It is widely used as a conductor of heat and electricity. There are several works about fiber-shaped devices built on copper wire. Through thermal treatment, CuO nanowires can be grown uniformly on a Cu wire. The insulating CuO is beneficial for suppressing the self-discharge in devices. After coating with MnO_2, a fiber electrode for supercapacitors can be readily prepared (Fig. 2.2c) [25].

Noble metal wires, such as Pt wire, Ag wire, and Au wire, are commonly used fiber electrodes [22, 26, 27]. For example, Pt is a common counter electrode in dye-sensitized solar cells and fuel cells because of its remarkable conductivity and catalytic activity to redox couple. In a plain junction solar cell, electrode with suitable work function is important for built-in electric field that drives the separation and transportation of photogenerated charges or excitons. Ag and Au are favored as high work function electrodes. However, these noble wires are expensive and less reserved in nature, and they are not as flexible to conform various curvatures as other fiber electrodes, which induces internal stress within the device.

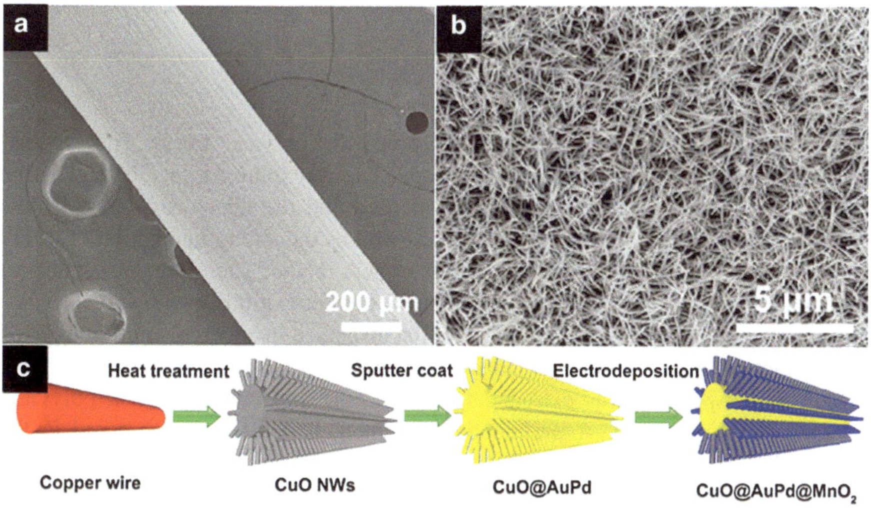

Fig. 2.2 **a** An SEM image of a Ni wire. **b** An SEM image of Co_3O_4 nanowires on a Ni fiber (Reproduced from Ref. [19] by permission of John Wiley & Sons Ltd). **c** The fabrication process of CuO@AuPd@MnO_2 nanowires (Reproduced from Ref. [25] by permission of John Wiley & Sons Ltd)

2.3 Carbon Nanotube Fiber

Carbon nanostructured materials including three-dimensional carbon sponges, two-dimensional graphene, one-dimensional carbon nanotubes, and zero-dimensional carbon spheres have been widely appreciated for their chemical and physical properties. Encouraging progress has been made to assemble carbon nanostructured materials like carbon nanotubes (CNTs) and graphene into macroscopic scale and extend their application in energy devices.

Carbon nanotubes have attracted a great deal of interest in academic and industry since 1991 [28]. Because of the excellent performance in electrical conductivity (10^5 S cm^{-1}), Young's modulus (0.9 TPa), tensile strength (150 GPa), thermal conductivity (3,500 W m^{-1} K^{-1}), and optical properties, CNTs have been broadly used in electrodes, actuators, filters, transistors, and structural fibers [29–31]. In order to translate the excellent properties of individual CNT into a large scale, it is necessary to fabricate macroscopic assemblies of CNTs like CNT fiber and CNT film [32]. The translation, however, is still challengeable. In this section, we will summarize the methods to prepare CNT fibers with an emphasis on aligned CNT fibers and discuss their properties.

2.3.1 Preparation Methods

Many methods have been developed to prepare CNT fiber through different precursors, as discussed in this section.

2.3.1.1 Wet Spinning

Wet spinning is a tradition method for preparing polymer fibers. This process requires a polymer solvent as spinning dope. The wet spinning gets its name from the wet bath where the spinneret is submerged and the fiber coagulates. Unlike polymer, the intrinsic chemical inertness as well as the strong interaction of CNTs limits their solubility in aqueous, organic, or acid media. Therefore, surfactants or chemical treatments are necessary to increase the compatibility with solvent.

For example, single-walled CNTs are first dispersed in aqueous solution assisted by surfactant. After sonication, black homogeneous suspensions were obtained. Then, the suspension was extruded from a needle and injected into a polyvinyl alcohol (PVA) solution. A preferential alignment of CNTs was observed in the obtained fiber arising from the flow-induced orientation during spinning. The tensile strength of CNT fiber was 150 MPa and conductivity was 10 S cm^{-1}. These CNT fibers exhibited a good flexibility and can be bent and tied without breaking [33]. Keeping PVA within the as-spun CNT fiber can significantly enhance the tensile strength to 1.8 GPa. Adding H_3PO_4 into the PVA, the CNT/PVA fiber can be used in supercapacitors where the H_3PO_4-contained PVA acted as electrolyte. However, it seemed to be unfeasible to prepare pure CNT fibers via this wet-spinning technique as it contains impurities like surfactant and polymers which is detrimental to the performance of CNTs especially the conductivity [34, 35].

Inspired from the previous results that single-walled CNT can be dissolved in sulfuric acid fumes and form liquid-crystalline phase, Ericson et al. prepared a single-walled CNT fiber from a CNT suspension in superacids [36]. CNTs in the fiber were found highly aligned. The single-walled CNT fibers possessed decent mechanical properties (Young's modulus of 120 GPa and a tensile strength of 116 MPa), high electrical conductivity (5,000 S cm^{-1}), and thermal conductivity (21 W K^{-1} m^{-1}) due to the absence of impurities. The processability of CNT solutions was improved by using chlorosulfonic acid to enhance the solubility of CNT in liquid phase. Controlling the liquid-crystalline phase behavior of single-walled CNTs in superacids, Davis et al. successfully prepared single-walled CNT fibers and films by flow-processing liquid-crystalline CNT acid dopes and quenching the fiber in coagulant [37]. The high concentrated CNT solutions favored the formation of liquid-crystal domains, and thus CNTs can be readily assembled into macroscopic fibers or sheets with controlled morphologies. Mechanical and electrical tests indicated that the obtained single-walled CNT fibers had a strength within 50 and 150 MPa and an electrical conductivity up to 8,300 S cm^{-1}.

These achievements encouraged the engineering of macroscopic single-walled CNT fibers through a superacid route, but it is not effective for multiwalled CNTs. Zhang et al. reported a coagulation process where multiwalled CNT fibers were spun from a liquid-crystalline ethylene glycol dispersion. This process allowed two types of CNTs: multiwalled CNTs and *N*-doped multiwalled CNTs. The modulus of *N*-doped CNT fibers was 142 ± 70 GPa, with a tensile strength of 0.17 ± 0.07 GPa. The wide range distribution of the modulus was attributed to the voids and defects generated in spinning process. The electrical conductivity of multiwalled CNT fibers is about 80 S cm^{-1}, while that of the *N*-doped multiwalled CNT fibers is about 300 S cm^{-1}. Further optimization of the spinning process is imperative to improve the mechanical and electrical properties of multiwalled CNT fibers [38]. Using longer CNTs (5 μm) and chlorosulfonic acid as solvent, Behabtu et al. reported a high-performance CNT fiber that had a high tensile strength (1.0 ± 0.2 GPa) and electrical conductivity ($2.9 \pm 0.3 \times 10^4$ S cm^{-1}, $5 \pm 0.5 \times 10^4$ S cm^{-1} after doping by iodine) [39]. In wet-spinning process, technique optimization and process control play a more important role in the resulting performance of CNT fiber, while the CNT characters including length, diameters, and wall numbers, notwithstanding some exceptions, are not dominant.

2.3.1.2 Synthesizing Ultralong Carbon Nanotubes

Long CNTs and stronger interactions are always in favor of sustaining the superior properties of individual CNTs in macroscopic scale. CNT fiber with longer CNTs has less CNT ends and inter-tube junctions, yielding higher strength and electrical

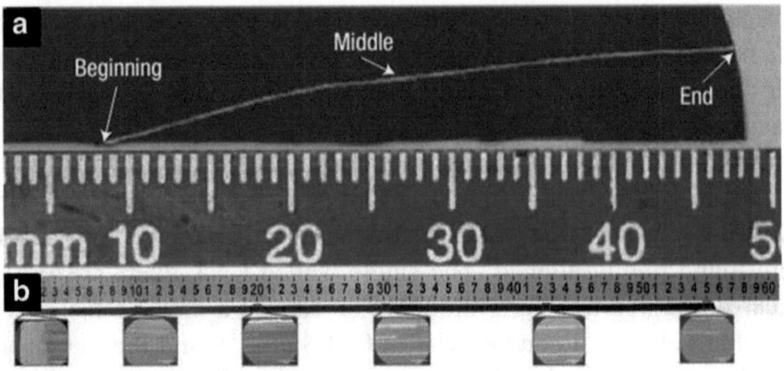

Fig. 2.3 Direct growth of an ultralong CNT on the substrate. **a** 4.6 cm (Reproduced from Ref. [40] by permission of Nature Publishing Group). **b** Preparation of a 55 cm of CNT (Reprinted with the permission from Ref. [41]. Copyright 2013 American Chemical Society)

conductivity. Zheng et al. synthesized 4 cm single-walled CNTs in an ambient condition furnace, which was afterwards increased to 55 cm after optimization (Fig. 2.3) [40, 41]. The tensile strength of the ultralong CNT was recorded 120 GPa with breaking strain of 15 %, which reached theoretical values. But the yield of ultralong CNT is too low to suffice for application.

2.3.1.3 Floating Chemical Vapor Deposition

In the effort to synthesize structure-controllable CNTs and their macro-assemblies, an encouraging breakthrough was made by Zhu et al. that macroscopic single-walled CNT fibers were prepared through a floating chemical vapor deposition (CVD) method [42]. The single-walled CNT fibers exhibited a metallic feature that their conductivity decreased from 2,000 to 1,400 S cm^{-1} with temperature rising from 90 to 300 K. The high conductivity of the fiber suggests that there were continuous conducting pathways in the fibers.

Li et al. prepared continuous CNT fibers by mechanically drawing CNT aerogel from the gaseous reaction zone and then directly winding it onto a roller (Fig. 2.4) [43]. This method relied on assembling CNTs in the gas flow via van der Waals interactions. Ethanol or acetone vapor are usually employed as carbon sources. The electrical conductivity of the as-received CNT fibers was 8,000 S cm^{-1}, and the tensile strength had a wide range between 0.10 and 1 GPa. They further advanced this approach and employed a mixture of ethanol and acetone vapor as the carbon source to fabricate continuous CNT fibers with a multilayered structure [44]. The continuous CNT fiber can be as long as several kilometers with conductivity of 5,000 S cm^{-1}. Besides, this process was in favor of CNTs with large diameter but few walls, and the interaction between neighboring nanotubes enhanced the fiber strength [45, 46].

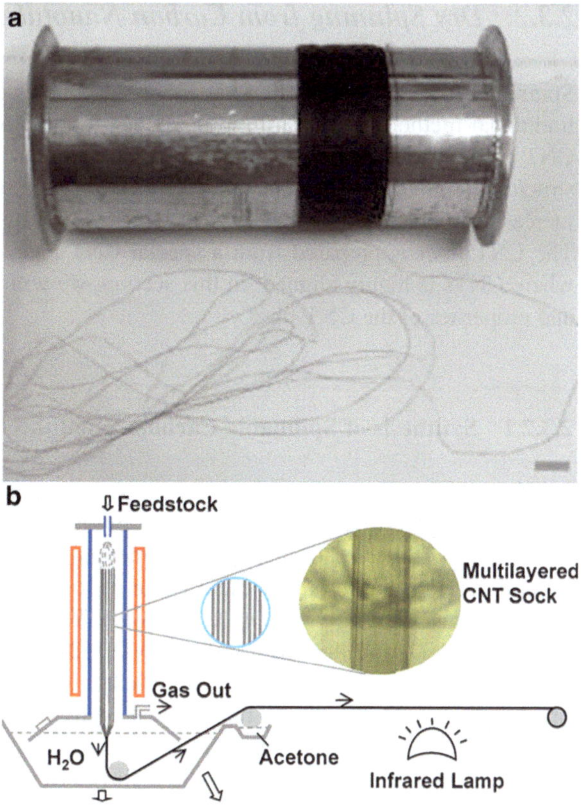

Fig. 2.4 **a** The CNT fiber prepared from CNT aerogel. **b** A scale-up production of continuous CNT fiber with multilayer structure (Reproduced from Ref. [44] by permission of John Wiley & Sons Ltd)

2.3.1.4 Dry Spinning

Twisting filaments into yarns is commonly used in industry to produce fiber materials like cotton thread. Accordingly, CNT fibers can be prepared in a similar approach by twisting the anisotropic CNT films, through which the fiber diameter and twisting angle can be controlled [47]. The CNT fibers exhibited crystal feature due to the highly aligned and densely packed CNTs inside the fiber. The conductivity of the fiber was close to 500 S cm^{-1} at room temperature [48]. Nevertheless, it should be pointed out that the length of the prepared fiber was limited by the CNT films, which is unfavorable for continuous production.

Inspired by the traditional technique of spinning fiber from pristine cotton, Ci et al. developed an efficient dry-spinning process to produce continuous double-walled CNT fibers from randomly CNT "cotton." Zheng et al. synthesized CNT cotton containing millimeters long CNTs by CVD approach and prepared CNT fibers thereafter. The mechanically robust as-spun fibers are applicable in efficient electron emission sources as well as electrodes for electrochemical sensing [49, 50]. Aligned CNT fibers that prove promising for fiber-shaped energy devices can be further dry spun from aligned CNT arrays and detailed in the following section as a widely studied procedure.

2.3.2 *Dry Spinning from Carbon Nanotube Array*

Spinning CNT fiber from CNT array is a new approach that exhibits advantages over traditional methods in preparing process and morphology control. The as-prepared CNT fiber, where CNTs are highly oriented, exhibits remarkable performances in structure, strength, and conductivity. Moreover, this aligned CNT fiber is gaining increasing popularity in energy devices which will be discussed in detail in this book. The CNT fiber is prepared from a special CNT precursor—spinnable CNT array, where CNTs is highly aligned. In this section, we will cast light on the preparation and properties of the CNT fiber.

2.3.2.1 Synthesis of Spinnable Carbon Nanotube Array

In 2002, Jiang et al. first reported preparing continuous CNT fibers from a spinnable CNT array. They tried to draw out a bundle of CNTs from a spinnable CNT array with several hundred micrometers in height [51]. To date, there are several methods based on chemical vapor deposition (CVD) to prepare spinnable CNT array including ambient condition CVD, low pressure CVD, plasma-enhanced CVD, and floating catalyst CVD [52–56]. According to catalyst preparation, these processes can be classified into one-step process and two-step process. In one-step process, the catalyst is dissolved in a precursor solution or deposited in a substrate [56, 57]. In a two-step process, the catalyst preparation is separated from the CNT growing which allows for a controllable CNT array with tailored catalysts [58]. The carbon sources supplying for spinnable CNT arrays mainly are methane, ethylene, acetylene, and cyclohexane. The growing temperature varies with different carbon sources. A higher temperature is in favor of defectless CNTs.

Here, a typical two-step preparation process using ethylene as carbon precursor is presented [59]. The catalyst is deposited on a silicon wafer with (100) crystal face. Not exclusively, the substrate can be expanded to any material with smooth surface and high-temperature resistivity like stainless steel. The catalyst exhibits a multilayered structure. Typically, a buffer layer (Al_2O_3, ZnO, MgO, etc.) is first deposited to prevent the penetration of catalyst into substrate and limit the mobility and Ostwald ripening of catalyst particles which is related to the diameter and wall numbers of the CNTs [60–62]. Above the buffer layer, a catalyst film (Fe, Co, Ni, or their alloy) is deposited and dictates the density and morphology of the resulting CNT array. For example, the thickness of the catalyst film affects the density as well as the diameter of formed nanoparticles. The CVD process is carried out in a furnace flowed with carbon source and carrier gas (H_2/Ar) with four programmed stages (Fig. 2.5a).

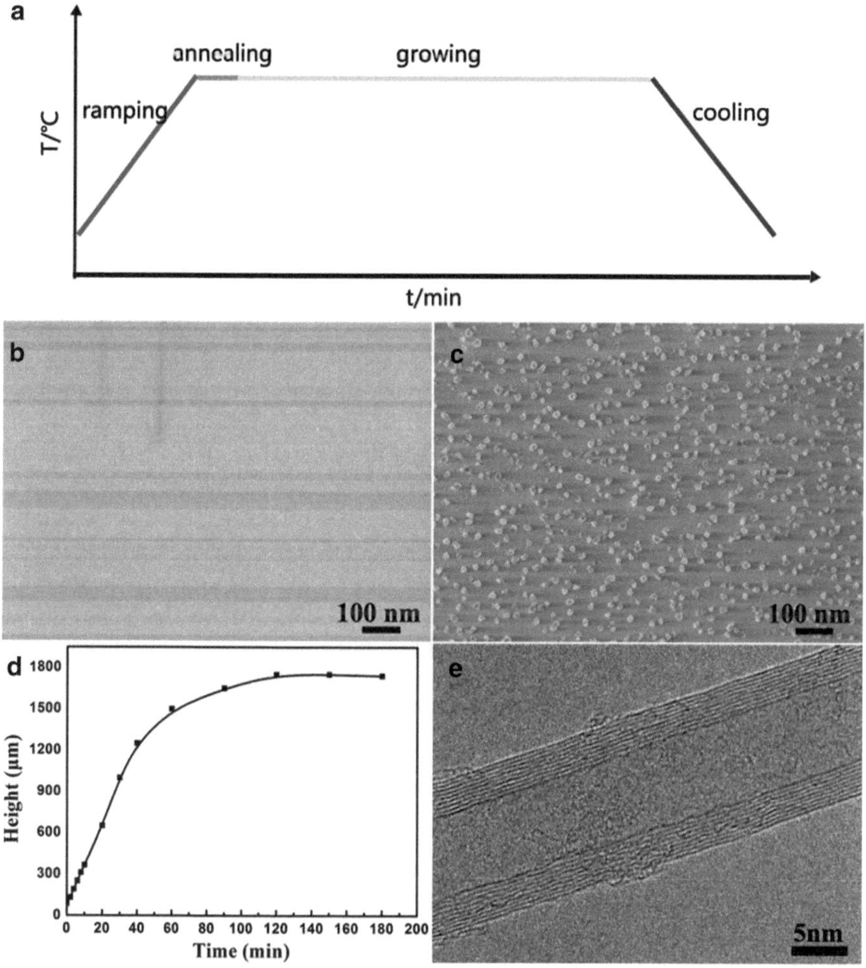

Fig. 2.5 **a** Programmed temperature for preparing CNT array through chemical vapor deposition. **b** An SEM image of the catalyst film morphology before annealing. **c** An SEM image of the catalyst nanoparticles formed after annealing. **d** The dependence of the height of CNT array on growing time. **e** A transmission electron microscopy (TEM) image of a multiwalled CNT in CNT array

2.3.2.2 Growing Mechanism

A representative growing mechanism of the CNT array conforms to the vapor–liquid–solid process [63]. At the first stage, the temperature ramps up and the catalyst film dewets into nanoparticles which are strewn over the buffer layer (Fig. 2.5b, c). The diameter and density of catalyst particles depend on the ramping rate, carrier gas component, compact buffer layer, and catalyst film thickness. As a single CNT grows from a single catalyst nanoparticle, the diameter and its distribution of the dewetted catalyst nanoparticles are vital for high-quality spinnable CNT arrays. Uniform catalyst nanoparticles always promise a narrow distribution of CNT's diameter. During the later process, carbon precursor, methane, ethylene, acetylene, or cyclohexane is carried into the high-temperature zone of furnace, where the carbon precursor is pyrolyzed and forms carbon clusters. The carbon clusters will dissolve into the catalyst nanoparticles and form iron carbide in a liquid or liquid-like state. Due to the high thermal conductivity and nanometer size of the catalyst as well as the rapid diffusion of carbon atoms, the temperature and the carbon atom concentration across the catalyst are uniform [63]. Once the catalyst is saturated, carbon atoms are prone to separate out and nucleate, which forms a carbon cap conforming to the shape of catalyst particles. The carbon caps can be single layered or multilayered, which leads to single-walled CNT or multiwalled CNT. Afterwards, a steady growth proceeds and the carbon caps are lifted up by the growing CNTs. The growing of CNT is accompanied by the formation of amorphous carbon that covers the catalyst particle and inhibits the carbon dissolution. Consequentially, the growing process terminates yielding short CNTs. Suppressing the amorphous carbon promises longer CNTs, for example, introducing a small amount of weak oxidant like water, ethanol into the gas flow helps to etch away the produced amorphous carbon and thus prolongs the growing process. Moreover, the weak oxidant also corrodes the outer walls of CNTs which favors CNTs with fewer even single wall [64, 65].

However, the longer CNTs do not promise spinnability. The growing condition of spinnable CNT array is narrowed within a small window arising from the rigorous requirements of morphology. A spinnable CNT array should have clean surface, oriented arrangement, and suitable density (Fig. 2.6a, b). Preparing spinnable CNT array starts with a delicately made catalyst whose components and thickness are precisely controlled. The morphology of CNT array strongly relies on the growing process involving gas components, flow rate, programmed temperature, and growing time. For example, the concentration of carbon source and H_2 is critical for competing reactions between amorphous carbon deposition and CNT growth. H_2 has an effect to refresh the iron catalyst and etch amorphous carbon. But as a pyrolysis product of carbon precursor, it also inhibits the balance towards decomposition. Temperature, of course, is another critical factor. Ordinarily, temperature is programmed for the ramp-up and hold stages. The ramp-up stage dictates the dewetting of catalyst particles, while the temperature plateau dominates the pyrolysis of carbon precursor as well as amorphous carbon deposition. The growth rate of CNT varies with catalyst size, which is determined by the oversaturation of carbon atoms. With different growth rates, the length distribution of CNT in the array will be broad and cause wavy structure inside which is detrimental for spinnability.

All of these factors contribute to the spinnability of CNTs. With a balance of flowing gas component, growing temperature, growing time, and suitable carbon source precursor, an amorphous carbon-free and highly aligned CNT array can be synthesized.

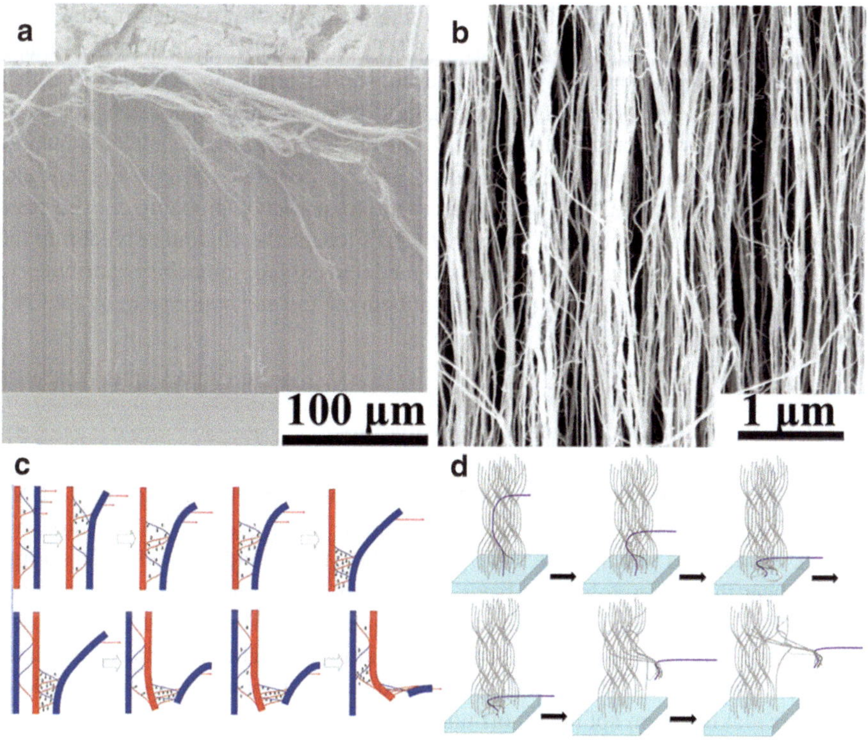

Fig. 2.6 SEM images of the spinnable CNT array and schematic illustrations of spinning mechanisms. **a** Side view. **b** High magnification images at middle **a. c** The interaction structure model. *Blue* and *red* strips represent CNT bundles. (Reprinted with the permission from Ref. [66]. Copyright 2011 American Chemical Society). **d** The entangled structure model. From Ref. [67] (Reprinted with permission from Elsevier)

2.3.2.3 Spinning Mechanism

It is the unique structure which has suitable interaction and entanglement among the neighboring CNT bundles that makes this CNT array spinnable. CNTs in the CNT array are apt to aggregate into bundles because of strong van der Waals interactions among tubes. In the dynamic CVD process, the gas flow causes disturbance and redistribution of CNT bundles, which leads to a unique structure that single CNT interpenetrates between two neighboring bundles. As shown in Fig. 2.6c, when a CNT bundle is being drawn out, CNTs penetrating the two neighboring bundles are divorced and gathered at the end of the bundles which suffice for peeling off the proximate CNT bundle [66]. Subsequently, CNT bundles are peeled off and drawn out continuously.

It is not hard to understand that the morphology of CNTs in the array dictates the spinnability. As the interaction between the CNTs is derived from the van der Waals force, a clean surface is naturally favorable. Amorphous carbon, however, affixed on

CNT surfaces as impurities, weakens these interactions so that it is detrimental to the spinnable CNT array [68]. The penetrated CNTs play a critical role during spinning, which increases with the CNT height. Hence, shorter arrays are reluctant to gather sufficient CNTs to involve the neighboring CNT bundles, whereas in the higher array, CNTs bundles are entangled into a bulk. On this account, the height of spinnable CNT array is balanced within a small range, which is controlled by growing time and flow rate. In some cases, CNTs in the array are found wavy, probably derived from turbulence in growing process. The wavy CNTs loosen the alignment rendering the CNT array hardly spinnable. Another used model to explain spinnability attributes to the entanglement formed during spinning at both the top and bottom part of the CNT array (Fig. 2.6d) [67, 69, 70].

2.3.2.4 Preparation of Carbon Nanotube Fiber

CNT fiber is usually derived from CNT sheet that is drawn from spinnable CNT array. There are basically two methods to assemble the aligned CNT sheet into CNT fibers. First one is twisting the CNT sheet [71]. The second method is densifying CNT sheet through a volatile solvent [54]. The twisting method produces a fiber where CNTs are packed at a helix angle with the fiber. The twisted CNTs have a large interaction and entanglement which promises a higher mechanical strength but a porous structure. The CNTs densified by solvent retain its orientation along the fiber making them apt to slip off under tension. On this account, a combination of twisting and densifying is more preferable in preparing CNT fibers. Moreover, displacing the solvent with the suspension of functional components during the twisting process, we can easily incorporate the guest components into the fiber [72]. In addition, applying a tensile stress during twisting by drawing fiber through a zigzag path where tensions are exerted between the pillars, the resulting CNT fibers are improved in alignment and increased in mechanical strength (Fig. 2.7c) [73].

Figure 2.7c shows the apparatus to prepare and collect CNT fibers. The spinnable CNT array is affixed on a rotation motor, and the other end of the CNT sheet drawn out is collected by a roller (Fig. 2.7a). When the motor rotates at a high speed, the CNT sheet is twisted in a typical triangle zone. The length of CNT fiber attainable is limited by the length of CNT array. Generally, a 1.5 mm length, several to tens of mm width, and around 250 μm height CNT array can be transferred into a 1 m fiber. The fiber diameter ranges within several to tens of micrometer which is determined by the width, thickness of CNT sheet, and rotation speed.

2.3.2.5 Properties

CNTs possess excellent mechanical, electrical, and thermal properties which are expected in CNT fibers. Admittedly, they are hardly achievable but approachable. The quality of the CNT array dictates the performance of the CNT fiber, including the CNT diameter, wall number, defects, and purity [75]. Besides, the preparing

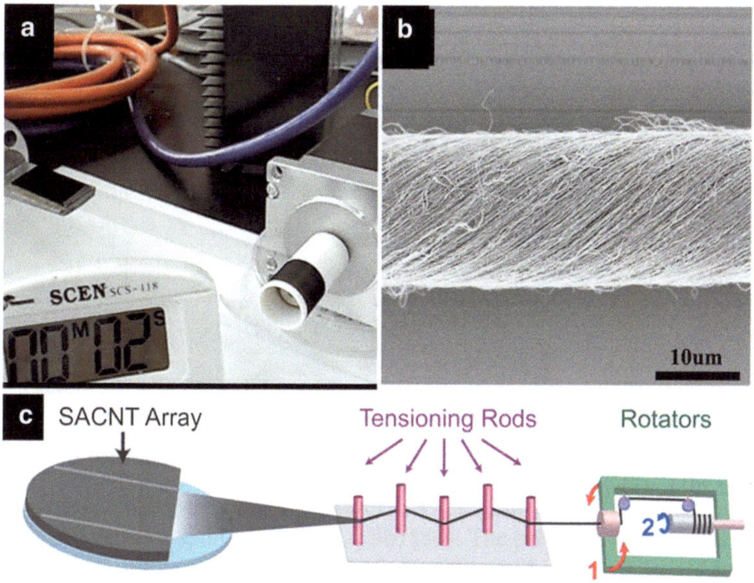

Fig. 2.7 Preparation of CNT fibers. **a** An aligned CNT sheet drawn from a spinnable CNT array. **b** An SEM image of a twisted CNT fiber. **c** A setup for continuous spinning process [74], with permission from Royal Society of Chemistry

process and posttreatment exert a strong impact. It is widely acknowledged in conventional textile industry that a twisting process is of the vital importance to the yarn properties. The general equation satisfying traditional fibers is introduced to describe the tensile strength σ_f of CNT fibers:

$$\sigma_f/\sigma_{\mathrm{CNT}} \approx \cos^2\vartheta_s(1 - k\operatorname{cosec}\vartheta_s) \tag{2.1}$$

where σ_{CNT}, ϑ_s, and k correspond to the tensile strength of CNT bundles, the surface twisting angle, and the load-transfer coefficient, respectively [76]. Obviously, the strength of CNT fiber strongly relies on its twisting angle, which can be controlled by the drawing speed and rotating speed. At its best, a CNT fiber has a tensile strength of 1.2 GPa [77]. When the fiber is over-twisted which induce the collapse of CNT, a 3.3 GPa of tensile strength was recorded [78].

Tracing the structural evolution in a typical stress–strain test is helpful to understand the fracture mechanism of the CNT fiber. In the beginning, the CNT bundle is densified under tension arising from the fact that the wavy CNTs are straightened. Afterwards, as the load increases, part of CNTs fractures and slips off, making the fiber loose. Then, the stress redistributes and concentrates on the intact CNT bundles; therefore, the CNT fiber tightened again (Fig. 2.8) [79]. Microstructure characterizations have demonstrated that the formed CNT fibers possess an end-to-end structure among nanotubes. Figure 2.9a illustrates two fibers composed of CNTs with different lengths. The end of a CNT is regarded as a defect because the intermolecular interaction at the CNT end is much weaker. In light of the fact

Co-spinning

Because of the porous structure in CNT fiber, the interspace between CNTs allows the invasion of functional guests during spinning. For example, immersing the CNT sheets into the suspension of nanoparticles, such as mesoporous carbon, and then twisting the CNT sheets, the nanoparticles are embraced within the resulting CNT fiber [90]. A continuous production can be proceeded on condition that suspension is dropped in the triangle zone in Fig. 2.7b, where the transformation of 2D sheet to 1D fiber takes places. Co-spinning is a feasible method for composition that takes advantage of the geometry transition from 2D to 1D during spinning. The 2D configuration of CNT sheet favors the deposition of functional components, which promises a high loading capacity. But due to the hydrophobicity of the CNTs, the co-spinning cannot carry out in aqueous suspension; thus, the dispersibility of functional materials in the solvent media should be considered.

Dip Coating

In some cases, a superficial deposition of materials is required. On this account, dip coating is more effective than co-spinning. As its name indicates, during this process, the CNT fiber is dipped into a solution and the interspaces in CNTs are infiltrated through capillarity. As the solvent evaporates, the material adheres to the surface of CNTs. As a demonstration, dipping CNT fibers into diacetylene monomer solution, they were coated with the monomer, and after polymerization, a CNT/polydiacetylene composite fiber was obtained, which has an electrochromatic behavior [83]. Many polymers are insoluble and hard to disperse, whereupon, dip coating provides a reliable strategy to incorporate polymers incompatible with conventional solvent [91, 92].

In Situ Compositing

In situ composition is a general description of a series of methods including in situ synthesis, electrochemical deposition, and electrochemical polymerization. For example, *N*-doped CNTs were synthesized at the outer surfaces of CNT fiber through chemical vapor deposition (Fig. 2.11a) [82]. Electrochemical depositing metal particles such as Ag or Pt can significantly enhance the conductivity and catalytic ability of the CNT fiber [94–96]. Polymerizing electrochemical active polymers like polyaniline and polypyrrole in their monomer solutions is a strategic approach to improve the capacity performance of the CNT fiber, which shows significance in fiber-shaped supercapacitors (Chap. 6) [97].

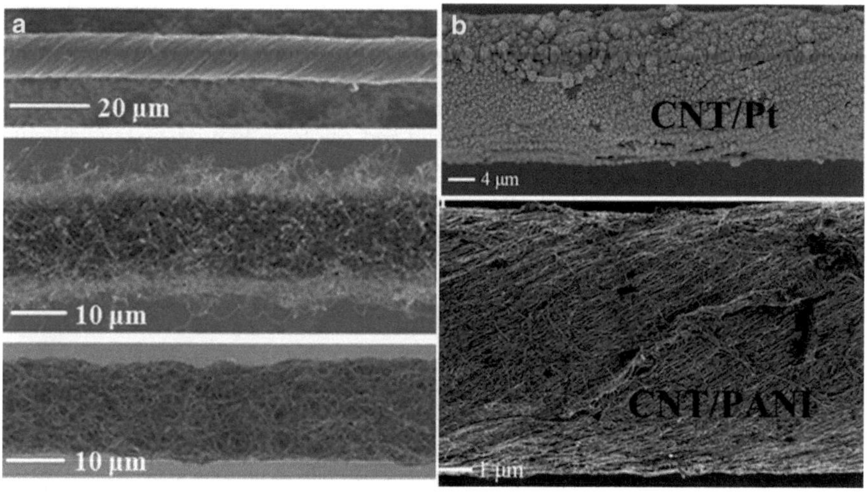

Fig. 2.11 CNT composite fibers prepared from in situ composition. **a** *N*-doped CNT fiber prepared from in situ synthesis (Reproduced from Ref. [82] by permission of John Wiley & Sons Ltd). **b** CNT/Pt and CNT/PANI composite fibers prepared through electrochemical deposition and polymerization (Reproduced from Ref. [93] by permission of John Wiley & Sons Ltd)

2.4 Graphene Fiber

Graphene, a single layer of sp^2-hybridized carbon atoms in a honeycomb lattice, is an arresting two-dimensional (2D) material with excellent mechanical, electrical, and thermal properties. It has a large theoretical specific surface area (2,630 $m^2\ g^{-1}$), high intrinsic mobility (200,000 $cm^2\ V^{-1}\ s^{-1}$), high Young's modulus (~1.0 TPa), high thermal conductivity (~5,000 $W\ m^{-1}\ K^{-1}$), and optical transmittance (~97.7 %) which make graphene applicable in many fields [98].

Nowadays, translating the properties of individual graphene sheets into macroscopic, ordered materials is necessary in view of exploring advanced properties of 2D graphene sheets for practical applications. In light of the achievements in preparing CNT fibers, assembling graphene sheets in a 1D fiber is naturally on the to-do list of researchers. Several approaches have been developed to prepare graphene fibers. Among them, wet-spinning method that extrudes the graphene oxide liquid crystal into a coagulation bath is reliable and widely used.

2.4.1 Synthesis

Making graphene fiber starts with graphene. Currently, there are several methods to prepare graphene: epitaxial growth, micromechanical exfoliation, and chemical oxidation. Epitaxial growth through chemical vapor deposition is ordinarily applied in preparing graphene with large size. However, this method is not suitable for continuous production. On the other hand, chemically oxidized graphene or graphene oxide (GO) can be synthesized at a large scale from graphite or CNT precursor. GO, the most accessible precursor of graphene, has good solubility in water, polar organic solvents, or super acid and can form lyotropic nematic liquid crystals. GO is usually synthesized through a modified Hummer method. Further oxidized, the solubility of GO increased. Other than graphite, CNTs also can be used as precursors for GO in ribbon structure [99].

Figure 2.12a, b show typical transmission electron microscopy (TEM) images of GO sheets. The GO sheets have widths of less than 2 mm and an average

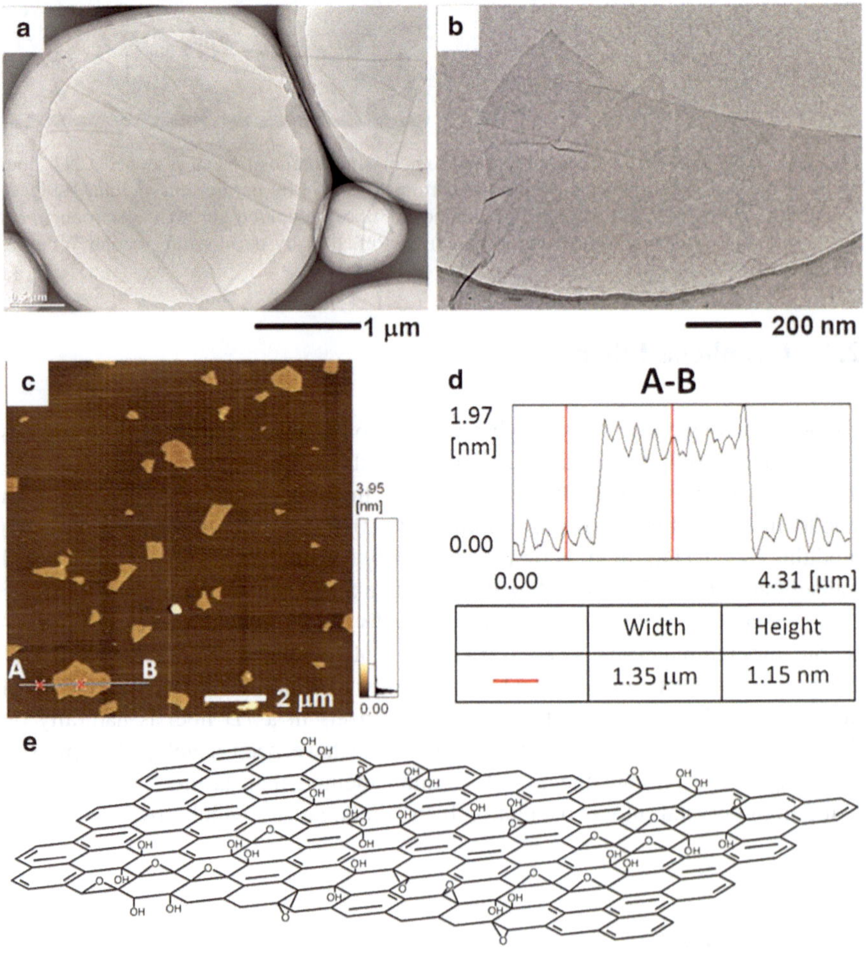

Fig. 2.12 Structure characterization of GO sheets. **a** and **b** TEM images of GO sheets at different magnifications. **c** An AFM image of GO sheets. **d** Height analysis of a GO sheet at **c** (Reprinted with the permission from Ref. [101]. Copyright 2013 John Wiley & Sons Ltd). **e** A proposed structure of GO (Reprinted by permission from Macmillan Publishers Ltd: Ref. [100], copyright 2009, with permission from Nature Publishing Group)

thickness of approximately 1.1 nm. Figure 2.12c shows a proposed structure of graphene oxide that is supported by solid-state nuclear magnetic resonance experiments on ^{13}C-labeled GO [100]. The oxidized graphite comprises stacked graphene layers which are scattered with oxygen-containing groups. The hydrophilicity of the dangling moieties makes water molecules readily intercalated into the graphene sheets. The interlayer distance between the GO sheets increases from 6 to 12 Å with increasing moisture. Graphene sheets in oxidized graphite can be completely exfoliated to produce aqueous colloidal suspensions assisted by sonication and mechanical agitation. The zeta-potential measurement manifests that the GO sheets are negatively charged in aqueous solution. The electrostatic repulsion between GO sheets prevents them against aggregation and stabilizes the suspension.

The size of GO exerts a significant impact on the mechanical and electrical properties of GO fibers. Pre-inflation of graphite at high temperature is inclined to produce large size GO sheets. Technically, gradient centrifugation can be adopted to get GO with uniform size. The smaller GO sheets are prone to disperse in supernate, while the larger ones will precipitate in the bottom. The GO aqueous suspension where GO forms lyotropic nematic liquid crystals is used as the raw material to produce GO fiber. Regularly arranged GO sheets in nematic liquid crystals promise the most viable fluid assembly approach towards macroscopic graphene fibers via wet-spinning process.

2.4.2 *Preparation*

As illustrated in Fig. 2.13a, the GO fiber is prepared by injecting GO suspension to a coagulation bath where GO sheets assemble into a continuous fiber. Then, the fiber is collected by a roller. The draw tension applied during collection can further align GO sheets in the fiber. The GO solution concentration, injecting speed, collecting speed, and nozzle diameter are contributable to the diameter of GO fiber. Subsequent chemical reduction, usually carried out in HI solution or hydrazine hydrate, imparts the resulting graphene fibers with high conductivity and good mechanical performance. After reduction, the interlayer space decreases to 0.37 nm, close to intrinsic space in graphite layers (0.34 nm) [102]. Figure 2.13b displays the surface morphology of the synthesized graphene fibers, which is affected by the component of the coagulation bath, involving the coagulation kinetics, mass transfer, and viscosity [103].

GO fiber can also be prepared through a dry-spinning process. As exemplified by Terrones et al., GO dispersions were first paved on a substrate, which were then dried into GO films. These films were subsequently scrolled into fibers with smooth surface, high toughness, outstanding ductility, and regular cross-sectional shape. The fiber spinning process was carried out in absence of water and can be used to composite with other functional components. When these fibers were over-twisted, they did not break. Instead, fibers self-coiled into a single fiber with a hierarchical twisting structure exhibiting the same chirality [104].

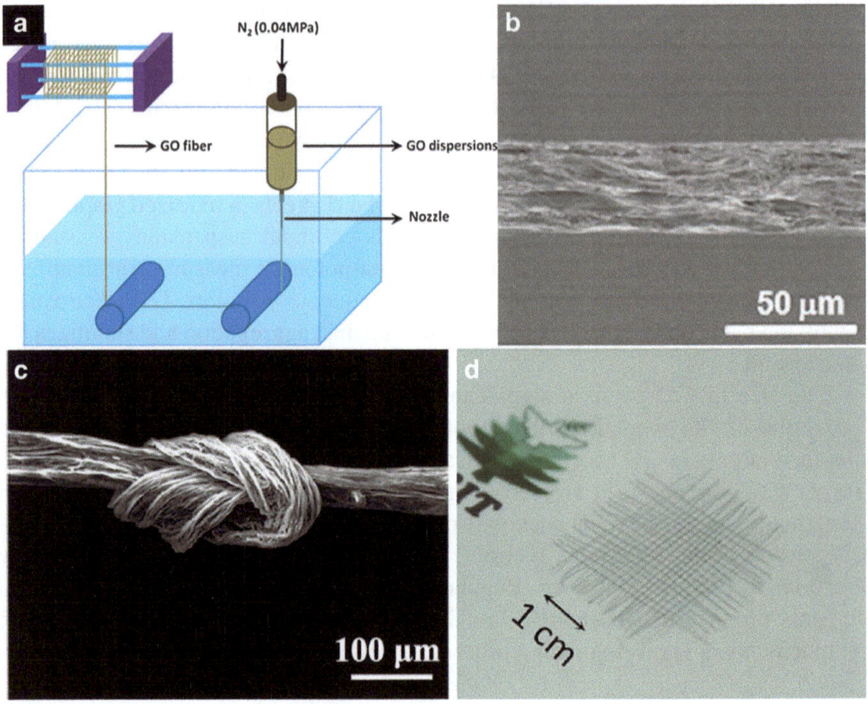

Fig. 2.13 **a** The preparation of GO fiber by wet-spinning method. **b** An SEM image of a GO fiber (Reprinted with the permission from Ref. [101]. Copyright 2013 John Wiley & Sons Ltd). **c** A knotted graphene fiber (Reproduced from Ref. [107] by permission of John Wiley & Sons Ltd). **d** A graphene fabric made from graphene fibers (Reproduced from Ref. [105] by permission of John Wiley & Sons Ltd)

Besides abovementioned methods, baking GO aqueous suspension in a glass tube is another method of making graphene fiber. The resulting GO fiber is conformable in the geometry of the tube and has a porous network structure, which allows introducing guest components during drying process [105].

Beside modified Hummer's method, there are other approaches for making graphene via solution process, e.g., graphene nanoribbons obtained from unzipping CNTs. Carretero et al. present a scalable method for fabricating long and narrow graphene nanoribbons that can be assembled and aligned into large graphene nanoribbon sheets and then into macroscopic graphene nanoribbon fibers with high conductivity and good mechanical performance [106].

2.4.3 Properties

GO fiber is composed of arranged GO sheets, and the interactions between sheets contain van der Waals force and hydrogen bond as there are many oxygen-containing groups. The rupture of GO fiber is caused by the slippage of neighboring sheets. Using larger GO sheets, higher strength can be achieved. Typically, after reduction the fiber exhibited a density of 0.61 g cm^{-3}, a strength of 10^2–10^3 MPa, and an electrical conductivity of 10^2–10^3 S cm^{-1} [101].

The produced graphene fiber exhibits excellent flexibility as shown in Fig. 2.13d. The graphene fiber can be knotted without breaking and weaved into a fabric. The properties of graphene fibers are affected by different spinning methods. For example, the GO fiber produced from wet spinning is fragile under deformation. Baking GO aqueous suspension in a glass tube provides a presetting structure of graphene fiber [105]. In a microscale, the performance of a graphene fiber is affected by the size and alignment of graphene sheets. Spinning from liquid-crystal phase can bring good intrinsic structural alignment within graphene fibers. A larger GO sheet may decrease the alignment but not for mechanical strength.

In light of this understanding, Gao et al. employed giant GO sheets with extremely high aspect ratios as building blocks to prepare graphene fibers with both high strength and conductivity. Besides, it has been proposed that divalent ions offer interlayer and intralayer cross-linking bridges between the oxygen-containing groups and thus bring the enhancement in the mechanical properties to GO bulk materials. With Ca^{2+} ions introduced, the tensile strength of GO fiber was elevated to 364.6 MPa. After chemical reduction, graphene fibers showed excellent electrical conductivities in the range of 3.8–4.1×10^4 S m^{-1} and mechanical strength to 501 MPa [108]. Tour et al. also demonstrated a dramatically improved mechanical performance arising from large flake GO fibers, with 178 % increase of specific stress, 188 % increase of specific modulus, and 278 % increase of elongation compared to the small flake GO fiber [107].

2.4.4 Composite Fiber

The performance of bare graphene fiber can be enhanced by introducing a secondary component to prepare a graphene composite fiber. There are several methods to prepare graphene composite fibers. For example, functional components can be introduced into the GO suspension and composited into the fiber through co-spinning. Yu et al. fabricated a graphene/poly(N-isopropylacrylamide) (PNIPAM) composite fiber by dissolving PNIPAM powders into the GO suspensions. The composite fiber exhibited a thermosensitive property inherited from PNIPAM [109].

Apart from polymer materials, inorganic functional nanomaterials can also be dispersed in GO suspension and extruded to prepare composite fiber. The performance of the composite fiber can be tailored by introducing different functional components. For example, a high conductive composite fiber can be obtained by introducing Ag particles into GO suspensions and co-spinning [110]. By introducing CNTs, the as-spun fiber shows enhanced mechanical strength and electrochemical capacitance [111].

Beneficial from the unique two-dimensional structure of graphene, which is favorable for nucleation and attachment, the secondary component can be straightly incorporated into graphene fiber. As a demonstration, the TiO_2 nanoparticles have been intercalated into the framework of graphene sheets by soaking the wet graphene fiber in commercial TiO_2 aqueous suspension. The TiO_2/graphene fiber displayed a photoresponse behavior benefited from the TiO_2, which is applicable in photodetectors, photocatalyst, and photovoltaic cells. Introducing different components, we can tailor the properties of the fiber. For example, introducing Fe_3O_4 nanoparticles can impart magnetism to composite fiber [105]. Depositing Pt nanoparticles on the graphene fiber can significantly enhance the electrocatalytic performance of the fiber which is conducive to broaden its application in electrochemical devices.

2.5 Carbon Fiber and Polymer Fiber

Besides the widely used metal wire and carbon nanomaterial-based conducting wire, there are many other flexible and functional fibers applicable in fiber-shaped energy devices.

2.5.1 Carbon Fiber

Carbon fiber is a potential material in electrocatalytic application. Carbon fiber has excellent mechanical and electrical properties with industry-scale production. Carbon fibers have high tensile strength up to 7 GPa with very good creep resistance, low densities ($\rho = 1.75–2.00$ g cm^{-3}), and high modulus up to 900 GPa. They are susceptible to oxidizing agents such as hot air and flames, but resistant to all other chemical species [112]. Polyacrylonitrile (PAN) and pitch are the two dominant precursors that produce carbon fibers with different properties. Bare carbon fiber can be used as counter electrode in fiber-shaped dye-sensitized solar cells. Introducing other functional materials can significantly enhance the performance. For example, coating photovoltaic polymers around carbon fiber, this flexible electrode shows promising application in fiber-shaped devices [113]. Beside electrochemical applications, carbon fiber can also serve as framework for functional materials. For example, Wang el al. deposited ZnO nanowires around carbon fiber using a hydrothermal method to prepare fiber-shaped electrodes [114]. ZnO nanowire-modified carbon fiber photoanode has been used in a dye-sensitized solar cell. Similarly, using dip-coating method, Zou et al. deposited TiO_2 nanoparticles around carbon fiber as fiber-shaped photoanode in an all-carbon solar cell [115].

2.5.2 Polymer Fiber

Polymer fibers are not fangled. In fact, we are wearing them every day. Like Nylon, Spandex, and polyester, polymer fibers take an important place in textile industries.

Produced from macromolecules, polymer fibers possess a series of merits including high mechanical strength, lightweight, flexibility, and chemical stability and can be imparted with various specialties via textile engineering. They should be favored by wearable devices but blocked by their intrinsic drawback: nonconductivity. Polymers like polyaniline, polypyrrole, and polythiophene have been found conductive or semiconductive in some circumstances, yet their fiber-shaped assemblies are still unavailable. On this account, electrically conductive polymer fiber seems a formidable challenge. Coating conducting materials such as conducting polymers, CNTs, and metal particles on the polymer fiber provides a feasible solution. For example, Nylon or PET fibers with 10–100 μm diameters can be coated with continuous thin films of the conducting polymer poly(3,4-ethylenedioxythiophene) (PEDOT) and used to create electrochemical transistor on single fibers [116]. The conductivity of the dip-coated conductive threads was less than 10 S cm^{-1}. The conductivity can be elevated to 100 S cm^{-1} by coating polystyrene sulfonic acid (PSS)-doped PEDOT. The conducting polymer-coated thread inherited the flexibility from the substrate and retained its performance under deformation. This strategy is compatible to various threads and conducting polymers [117].

Besides conducting polymer, other conducting materials like CNTs and graphene sheets are adopted to composite with polymer fibers. For example, by coating CNT suspension, we can paint a circuit on the cotton cloth [118]. Winding aligned CNT sheets around the fiber substrate is another approach towards conducting fibers. The as-prepared composite fiber exhibited a high conductivity (10^2–10^3 S m^{-1}) derived from the wrapped CNTs [119]. Moreover, displacing the polymer fiber with an elastic rubbery fiber, the composite fiber displays stretchable capability, which is applicable in wearable devices [120].

Other than coating a conducting cladding outside the polymer fiber, conducting additives can also be introduced during fiber spinning. For example, Ag nanoparticles can endow composite fiber with excellent conductivity, while CNT may enhance the charge transport and tensile strength. Most important, polymer component make the composite fiber highly stretchable. By combining these three parts into one single fiber using a wet-spinning method, Baik et al. successfully fabricated a highly conductive and stretchable product (17,460 S/cm with a rupture tensile strain of 50 %) [121].

2.6 Summary

It is not an exaggeration that fiber electrodes reign the fiber-shaped energy devices. As required by energy devices, high mechanical strength and electrical conductivity are prerequisite conditions for fiber electrode. In this chapter, electrically conducting fibers made from various materials are presented. The commonly used metal wires provide sufficient conductivity but are circumscribed by their heavy weight, high expenses, and limited flexibility. As the nanotechnology swept, fibers made from nanomaterials emerged and showed their advantages over the traditional materials. As excellent representatives, CNT fiber especially the aligned CNT fiber spun from the CNT array and graphene fiber is highlighted. They have remarkable intrinsic properties applicable as fiber electrodes, and they are also accessible for extrinsic components adaptable for different requirements, which make them play a pivotal role in fiber-shaped devices.

References

1. Hardin BE, Snaith HJ, McGehee MD (2012) The renaissance of dye-sensitized solar cells. Nat Photonics 6(3):162–169
2. Dou LT, You JB, Hong ZR, Xu Z, Li G, Street RA, Yang Y (2013) 25th anniversary article: a decade of organic/polymeric photovoltaic research. Adv Mater 25(46):6642–6671
3. Qiu L, Wu Q, Yang Z, Sun X, Zhang Y, Peng H (2014) Freestanding aligned carbon nanotube array grown on a large-area single-layered graphene sheet for efficient dye-sensitized solar cell. Small. doi:10.1002/smll.201400703:n/a-n/a
4. Chen T, Yang ZB, Peng HS (2013) Integrated devices to realize energy conversion and storage simultaneously. ChemPhysChem 14(9):1777–1782
5. Chen T, Qiu LB, Yang ZB, Peng HS (2013) Novel solar cells in a wire format. Chem Soc Rev 42(12):5031–5041
6. Jost K, Dion G, Gogotsi Y (2014) Textile energy storage in perspective. J Mater Chem A 2(28):10776–10787
7. Thomas S, Deepak TG, Anjusree GS, Arun TA, Nair SV, Nair AS (2014) A review on counter electrode materials in dye-sensitized solar cells. J Mater Chem A 2(13):4474–4490
8. Chen X, Mao SS (2007) Titanium dioxide nanomaterials: synthesis, properties, modifications, and applications. Chem Rev 107(7):2891–2959
9. Chen L, Zhou Y, Dai H, Li ZD, Yu T, Liu JG, Zou ZG (2013) Fiber dye-sensitized solar cells consisting of TiO_2 nanowires arrays on Ti thread as photoanodes through a low-cost, scalable route. J Mater Chem A 1(38):11790–11794
10. Lv ZB, Fu YP, Hou SC, Wang D, Wu HW, Zhang C, Chu ZZ, Zou DC (2011) Large size, high efficiency fiber-shaped dye-sensitized solar cells. Phys Chem Chem Phys 13(21):10076–10083
11. Hou SC, Cai X, Wu HW, Yu X, Peng M, Yan K, Zou DC (2013) Nitrogen-doped graphene for dye-sensitized solar cells and the role of nitrogen states in triiodide reduction. Energy Environ Sci 6(11):3356–3362
12. Zhang S, Ji CY, Bian ZQ, Liu RH, Xia XY, Yun DQ, Zhang LH, Huang CH, Cao AY (2011) Single-wire dye-sensitized solar cells wrapped by carbon nanotube film electrodes. Nano Lett 11(8):3383–3387
13. Chen T, Qiu LB, Kia HG, Yang ZB, Peng HS (2012) Designing aligned inorganic nanotubes at the electrode interface: towards highly efficient photovoltaic wires. Adv Mater 24(34):4623–4628
14. Chen T, Qiu LB, Yang ZB, Cai ZB, Ren J, Li HP, Lin HJ, Sun XM, Peng HS (2012) An integrated "energy wire" for both photoelectric conversion and energy storage. Angew Chem Int Ed 51(48):11977–11980
15. Ramier J, Da Costa N, Plummer CJG, Leterrier Y, Manson JAE, Eckert R, Gaudiana R (2008) Cohesion and adhesion of nanoporous TiO_2 coatings on titanium wires for photovoltaic applications. Thin Solid Films 516(8):1913–1919
16. Docampo P, Guldin S, Leijtens T, Noel NK, Steiner U, Snaith HJ (2014) Lessons learned: from dye-sensitized solar cells to all-solid-state hybrid devices. Adv Mater 26(24):4013–4030
17. Grimes CA (2007) Synthesis and application of highly ordered arrays of TiO_2 nanotubes. J Mater Chem 17(15):1451–1457

18. Wang H, Liu Y, Li M, Huang H, Zhong MY, Shen H (2009) Hydrothermal growth of large-scale macroporous TiO_2 nanowires and its application in 3D dye-sensitized solar cells. Appl Phys Mater Sci Process 97(1):25–29
19. Wang XF, Liu B, Liu R, Wang QF, Hou XJ, Chen D, Wang RM, Shen GZ (2014) Fiber-based flexible all-solid-state asymmetric supercapacitors for integrated photodetecting system. Angew Chem Int Ed 53(7):1849–1853
20. Lee MR, Eckert RD, Forberich K, Dennler G, Brabec CJ, Gaudiana RA (2009) Solar power wires based on organic photovoltaic materials. Science 324(5924):232–235
21. Fan X, Chu ZZ, Wang FZ, Zhang C, Chen L, Tang YW, Zou DC (2008) Wire-shaped flexible dye-sensitized solar cells. Adv Mater 20(3):592–595
22. Liu DY, Zhao MY, Li Y, Bian ZQ, Zhang LH, Shang YY, Xia XY, Zhang S, Yun DQ, Liu ZW, Cao AY, Huang CH (2012) Solid-state, polymer-based fiber solar cells with carbon nanotube electrodes. ACS Nano 6(12):11027–11034
23. Fu YP, Lv ZB, Hou SC, Wu HW, Wang D, Zhang C, Zou DC (2012) TCO-free, flexible, and bifacial dye-sensitized solar cell based on low-cost metal wires. Adv Energy Mater 2(1):37–41
24. Dong X, Guo Z, Song Y, Hou M, Wang J, Wang Y, Xia Y (2014) Flexible and wire-shaped micro-supercapacitor based on Ni(OH)2-nanowire and ordered mesoporous carbon electrodes. Adv Funct Mater 24(22):3405–3412
25. Yu ZA, Thomas J (2014) Energy storing electrical cables: integrating energy storage and electrical conduction. Adv Mater 26(25):4279–4285
26. Huang SQ, Guo XZ, Huang XM, Zhang QX, Sun HC, Li DM, Luo YH, Meng QB (2011) Highly efficient fibrous dye-sensitized solar cells based on TiO_2 nanotube arrays. Nanotechnology 22(31):315402
27. Wang D, Hou SC, Wu HW, Zhang C, Chu ZZ, Zou DC (2011) Fiber-shaped all-solid state dye sensitized solar cell with remarkably enhanced performance via substrate surface engineering and TiO_2 film modification. J Mater Chem 21(17):6383–6388
28. Iijima S (1991) Helical microtubules of graphitic carbon. Nature 354(6348):56–58
29. Ebbesen TW, Lezec HJ, Hiura H, Bennett JW, Ghaemi HF, Thio T (1996) Electrical conductivity of individual carbon nanotubes. Nature 382(6586):54–56
30. Yu MF, Files BS, Arepalli S, Ruoff RS (2000) Tensile loading of ropes of single wall carbon nanotubes and their mechanical properties. Phys Rev Lett 84(24):5552–5555
31. Kim P, Shi L, Majumdar A, McEuen PL (2001) Thermal transport measurements of individual multiwalled nanotubes. Phys Rev Lett 87(21):215502
32. Liu LQ, Ma WJ, Zhang Z (2011) Macroscopic carbon nanotube assemblies: preparation, properties, and potential applications. Small 7(11):1504–1520
33. Vigolo B, Penicaud A, Coulon C, Sauder C, Pailler R, Journet C, Bernier P, Poulin P (2000) Macroscopic fibers and ribbons of oriented carbon nanotubes. Science 290(5495):1331–1334
34. Dalton AB, Collins S, Munoz E, Razal JM, Ebron VH, Ferraris JP, Coleman JN, Kim BG, Baughman RH (2003) Super-tough carbon-nanotube fibres – these extraordinary composite fibres can be woven into electronic textiles. Nature 423(6941):703–703
35. Dalton AB, Collins S, Razal J, Munoz E, Ebron VH, Kim BG, Coleman JN, Ferraris JP, Baughman RH (2004) Continuous carbon nanotube composite fibers: properties, potential applications, and problems. J Mater Chem 14(1):1–3
36. Ericson LM, Fan H, Peng HQ, Davis VA, Zhou W, Sulpizio J, Wang YH, Booker R, Vavro J, Guthy C, Parra-Vasquez ANG, Kim MJ, Ramesh S, Saini RK, Kittrell C, Lavin G, Schmidt H, Adams WW, Billups WE, Pasquali M, Hwang WF, Hauge RH, Fischer JE, Smalley RE (2004) Macroscopic, neat, single-walled carbon nanotube fibers. Science 305 (5689):1447–1450
37. Davis VA, Parra-Vasquez ANG, Green MJ, Rai PK, Behabtu N, Prieto V, Booker RD, Schmidt J, Kesselman E, Zhou W, Fan H, Adams WW, Hauge RH, Fischer JE, Cohen Y, Talmon Y, Smalley RE, Pasquali M (2009) True solutions of single-walled carbon nanotubes for assembly into macroscopic materials. Nat Nanotechnol 4(12):830–834

38. Zhang SJ, Koziol KKK, Kinloch IA, Windle AH (2008) Macroscopic fibers of well-aligned carbon nanotubes by wet spinning. Small 4(8):1217–1222
39. Behabtu N, Young CC, Tsentalovich DE, Kleinerman O, Wang X, Ma AWK, Bengio EA, ter Waarbeek RF, de Jong JJ, Hoogerwerf RE, Fairchild SB, Ferguson JB, Maruyama B, Kono J, Talmon Y, Cohen Y, Otto MJ, Pasquali M (2013) Strong, light, multifunctional fibers of carbon nanotubes with ultrahigh conductivity. Science 339(6116):182–186
40. Zheng LX, O'Connell MJ, Doorn SK, Liao XZ, Zhao YH, Akhadov EA, Hoffbauer MA, Roop BJ, Jia QX, Dye RC, Peterson DE, Huang SM, Liu J, Zhu YT (2004) Ultralong single-wall carbon nanotubes. Nat Mater 3(10):673–676
41. Zhang RF, Zhang YY, Zhang Q, Xie HH, Qian WZ, Wei F (2013) Growth of half-meter long carbon nanotubes based on Schulz-Flory distribution. ACS Nano 7(7):6156–6161
42. Zhu HW, Xu CL, Wu DH, Wei BQ, Vajtai R, Ajayan PM (2002) Direct synthesis of long single-walled carbon nanotube strands. Science 296(5569):884–886
43. Li YL, Kinloch IA, Windle AH (2004) Direct spinning of carbon nanotube fibers from chemical vapor deposition synthesis. Science 304(5668):276–278
44. Zhong XH, Li YL, Liu YK, Qiao XH, Feng Y, Liang J, Jin J, Zhu L, Hou F, Li JY (2010) Continuous multilayered carbon nanotube yarns. Adv Mater 22(6):692–696
45. Motta M, Moisala A, Kinloch IA, Windle AH (2007) High performance fibres from 'Dog bone' carbon nanotubes. Adv Mater 19(21):3721–3726
46. Koziol K, Vilatela J, Moisala A, Motta M, Cunniff P, Sennett M, Windle A (2007) High-performance carbon nanotube fiber. Science 318(5858):1892–1895
47. Ma WJ, Liu LQ, Yang R, Zhang TH, Zhang Z, Song L, Ren Y, Shen J, Niu ZQ, Zhou WY, Xie SS (2009) Monitoring a micromechanical process in macroscale carbon nanotube films and fibers. Adv Mater 21(5):603–608
48. Liu GT, Zhao YC, Deng K, Liu Z, Chu WG, Chen JR, Yang YL, Zheng KH, Huang HB, Ma WJ, Song L, Yang HF, Gu CZ, Rao GH, Wang C, Xie SS, Sun LF (2008) Highly dense and perfectly aligned single-walled carbon nanotubes fabricated by diamond wire drawing dies. Nano Lett 8(4):1071–1075
49. Ci L, Punbusayakul N, Wei JQ, Vajtai R, Talapatra S, Ajayan PM (2007) Multifunctional macroarchitectures of double-walled carbon nanotube fibers. Adv Mater 19(13):1719–1723
50. Zheng LX, Zhang XF, Li QW, Chikkannanavar SB, Li Y, Zhao YH, Liao XZ, Jia QX, Doorn SK, Peterson DE, Zhu YT (2007) Carbon-nanotube cotton for large-scale fibers. Adv Mater 19(18):2567–2570
51. Jiang KL, Li QQ, Fan SS (2002) Nanotechnology: spinning continuous carbon nanotube yarns – Carbon nanotubes weave their way into a range of imaginative macroscopic applications. Nature 419(6909):801–801
52. Zhang M, Fang SL, Zakhidov AA, Lee SB, Aliev AE, Williams CD, Atkinson KR, Baughman RH (2005) Strong, transparent, multifunctional, carbon nanotube sheets. Science 309(5738):1215–1219
53. Huynh CP, Hawkins SC (2010) Understanding the synthesis of directly spinnable carbon nanotube forests. Carbon 48(4):1105–1115
54. Zhang XB, Jiang KL, Teng C, Liu P, Zhang L, Kong J, Zhang TH, Li QQ, Fan SS (2006) Spinning and processing continuous yarns from 4-inch wafer scale super-aligned carbon nanotube arrays. Adv Mater 18(12):1505–1510
55. Ryu S, Lee Y, Hwang JW, Hong S, Kim C, Park TG, Lee H, Hong SH (2011) High-strength carbon nanotube fibers fabricated by infiltration and curing of mussel-inspired catecholamine polymer. Adv Mater 23(17):1971–1975
56. Zhang Q, Wang DG, Huang JQ, Zhou WP, Luo GH, Qian WZ, Wei F (2010) Dry spinning yarns from vertically aligned carbon nanotube arrays produced by an improved floating catalyst chemical vapor deposition method. Carbon 48(10):2855–2861
57. Inoue Y, Kakihata K, Hirono Y, Horie T, Ishida A, Mimura H (2008) One-step grown aligned bulk carbon nanotubes by chloride mediated chemical vapor deposition. Appl Phys Lett 92(21):213113

58. Kim JH, Jang HS, Lee KH, Overzet LJ, Lee GS (2010) Tuning of Fe catalysts for growth of spin-capable carbon nanotubes. Carbon 48(2):538–547
59. Qiu LB, Sun XM, Yang ZB, Guo WH, Peng HS (2012) Preparation and application of aligned carbon nanotube/polymer composite material. Acta Chim Sin 70(14):1523–1532
60. Amama PB, Pint CL, Kim SM, McJilton L, Eyink KG, Stach EA, Hauge RH, Maruyama B (2010) Influence of alumina type on the evolution and activity of alumina-supported Fe catalysts in single-walled carbon nanotube carpet growth. ACS Nano 4(2):895–904
61. Jia JJ, Zhao JN, Xu G, Di JT, Yong ZZ, Tao YY, Fang CO, Zhang ZG, Zhang XH, Zheng LX, Li QW (2011) A comparison of the mechanical properties of fibers spun from different carbon nanotubes. Carbon 49(4):1333–1339
62. Zhang YY, Zou GF, Doorn SK, Htoon H, Stan L, Hawley ME, Sheehan CJ, Zhu YT, Jia QX (2009) Tailoring the morphology of carbon nanotube arrays: from spinnable forests to undulating foams. ACS Nano 3(8):2157–2162
63. Jiang KL, Feng C, Liu K, Fan SS (2007) A vapor-liquid-solid model for chemical vapor deposition growth of carbon nanotubes. J Nanosci Nanotechnol 7(4–5):1494–1504
64. Hata K, Futaba DN, Mizuno K, Namai T, Yumura M, Iijima S (2005) Water-assisted highly efficient synthesis of impurity-free single-walled carbon nanotubes. Abstr Pap Am Chem Soc 229:U967–U967
65. Zhang S, Zhu L, Minus ML, Chae HG, Jagannathan S, Wong CP, Kowalik J, Roberson LB, Kumar S (2008) Solid-state spun fibers and yarns from 1-mm long carbon nanotube forests synthesized by water-assisted chemical vapor deposition. J Mater Sci 43(13):4356–4362
66. Kuznetsov AA, Fonseca AF, Baughman RH, Zakhidov AA (2011) Structural model for dry-drawing of sheets and yarns from carbon nanotube forests. ACS Nano 5(2):985–993
67. Zhu C, Cheng C, He YH, Wang L, Wong TL, Fung KK, Wang N (2011) A self-entanglement mechanism for continuous pulling of carbon nanotube yarns. Carbon 49(15):4996–5001
68. Liu K, Sun YH, Liu P, Wang JP, Li QQ, Fan SS, Jiang KL (2009) Periodically striped films produced from super-aligned carbon nanotube arrays. Nanotechnology 20(33):335705
69. Gilvaei AF, Hirahara K, Nakayama Y (2011) In-situ study of the carbon nanotube yarn drawing process. Carbon 49(14):4928–4935
70. Wei HM, Wei Y, Wu Y, Liu L, Fan SS, Jiang KL (2013) High-strength composite yarns derived from oxygen plasma modified super-aligned carbon nanotube arrays. Nano Res 6 (3):208–215
71. Zhang M, Atkinson KR, Baughman RH (2004) Multifunctional carbon nanotube yarns by downsizing an ancient technology. Science 306(5700):1358–1361
72. Liu K, Sun YH, Zhou RF, Zhu HY, Wang JP, Liu L, Fan SS, Jiang KL (2010) Carbon nanotube yarns with high tensile strength made by a twisting and shrinking method. Nanotechnology 21(4):045708
73. Tran CD, Humphries W, Smith SM, Huynh C, Lucas S (2009) Improving the tensile strength of carbon nanotube spun yarns using a modified spinning process. Carbon 47(11):2662–2670
74. Liu K, Zhu F, Liu L, Sun YH, Fan SS, Jiang KL (2012) Fabrication and processing of high-strength densely packed carbon nanotube yarns without solution processes. Nanoscale 4(11):3389–3393
75. Liu K, Sun YH, Chen L, Feng C, Feng XF, Jiang KL, Zhao YG, Fan SS (2008) Controlled growth of super-aligned carbon nanotube arrays for spinning continuous unidirectional sheets with tunable physical properties. Nano Lett 8(2):700–705
76. Beyerlein IJ, Porwal PK, Zhu YT, Hu K, Xu XF (2009) Scale and twist effects on the strength of nanostructured yarns and reinforced composites. Nanotechnology 20(48):485702
77. Zhao JN, Zhang XH, Di JT, Xu G, Yang XJ, Liu XY, Yong ZZ, Chen MH, Li QW (2010) Double-peak mechanical properties of carbon-nanotube fibers. Small 6(22):2612–2617
78. Zhang XF, Li QW, Holesinger TG, Arendt PN, Huang JY, Kirven PD, Clapp TG, DePaula RF, Liao XZ, Zhao YH, Zheng LX, Peterson DE, Zhu YT (2007) Ultrastrong, stiff, and lightweight carbon-nanotube fibers. Adv Mater 19(23):4198–4201

79. Deng F, Lu WB, Zhao HB, Zhu YT, Kim BS, Chou TW (2011) The properties of dry-spun carbon nanotube fibers and their interfacial shear strength in an epoxy composite. Carbon 49 (5):1752–1757
80. Behabtu N, Green MJ, Pasquali M (2008) Carbon nanotube-based neat fibers. Nano Today 3 (5–6):24–34
81. Zhang XF, Li QW, Tu Y, Li YA, Coulter JY, Zheng LX, Zhao YH, Jia QX, Peterson DE, Zhu YT (2007) Strong carbon-nanotube fibers spun from long carbon-nanotube arrays. Small 3 (2):244–248
82. Chen T, Cai ZB, Yang ZB, Li L, Sun XM, Huang T, Yu AS, Kia HG, Peng HS (2011) Nitrogen-doped carbon nanotube composite fiber with a core-sheath structure for novel electrodes. Adv Mater 23(40):4620–4625
83. Peng HS, Sun XM, Cai FJ, Chen XL, Zhu YC, Liao GP, Chen DY, Li QW, Lu YF, Zhu YT, Jia QX (2009) Electrochromatic carbon nanotube/polydiacetylene nanocomposite fibres. Nat Nanotechnol 4(11):738–741
84. Banerjee S, Chakravorty D (1998) Electrical resistivity of copper-silica nanocomposites synthesized by electrodeposition. J Appl Phys 84(2):1149–1151
85. Ma YG, Liu HJ, Ong CK (2006) Electron transport properties in CoAlO composite antidot arrays. Europhys Lett 76(6):1144–1150
86. Yang ZB, Sun XM, Chen XL, Yong ZZ, Xu G, He RX, An ZH, Li QW, Peng HS (2011) Dependence of structures and properties of carbon nanotube fibers on heating treatment. J Mater Chem 21(36):13772–13775
87. Meng FC, Zhao JN, Ye YT, Zhang XH, Li QW (2012) Carbon nanotube fibers for electrochemical applications: effect of enhanced interfaces by an acid treatment. Nanoscale 4(23):7464–7468
88. Lepro X, Ovalle-Robles R, Lima MD, Elias AL, Terrones M, Baughman RH (2012) Catalytic twist-spun yarns of nitrogen-doped carbon nanotubes. Adv Funct Mater 22(5):1069–1075
89. Sun H, Yang ZB, Chen XL, Qiu LB, You X, Chen PN, Peng HS (2013) Photovoltaic wire with high efficiency attached onto and detached from a substrate using a magnetic field. Angew Chem Int Ed 52(32):8276–8280
90. Lima MD, Fang SL, Lepro X, Lewis C, Ovalle-Robles R, Carretero-Gonzalez J, Castillo-Martinez E, Kozlov ME, Oh JY, Rawat N, Haines CS, Haque MH, Aare V, Stoughton S, Zakhidov AA, Baughman RH (2011) Biscrolling nanotube sheets and functional guests into yarns. Science 331(6013):51–55
91. Guo WH, Liu C, Sun XM, Yang ZB, Kia HG, Peng HS (2012) Aligned carbon nanotube/polymer composite fibers with improved mechanical strength and electrical conductivity. J Mater Chem 22(3):903–908
92. Li S, Zhang XH, Zhao JN, Meng FC, Xu G, Yong ZZ, Jia JJ, Zhang ZG, Li QW (2012) Enhancement of carbon nanotube fibres using different solvents and polymers. Compos Sci Technol 72(12):1402–1407
93. Chen T, Cai ZB, Qiu LB, Li HP, Ren J, Lin HJ, Yang ZB, Sun XM, Peng HS (2013) Synthesis of aligned carbon nanotube composite fibers with high performances by electrochemical deposition. J Mater Chem A 1(6):2211–2216
94. Randeniya LK, Bendavid A, Martin PJ, Tran CD (2010) Composite yarns of multiwalled carbon nanotubes with metallic electrical conductivity. Small 6(16):1806–1811
95. Xu G, Zhao JN, Li S, Zhang XH, Yong ZZ, Li QW (2011) Continuous electrodeposition for lightweight, highly conducting and strong carbon nanotube-copper composite fibers. Nanoscale 3(10):4215–4219
96. Zhang S, Ji CY, Bian ZQ, Yu PR, Zhang LH, Liu DY, Shi EZ, Shang YY, Peng HT, Cheng Q, Wang D, Huang CH, Cao AY (2012) Porous, platinum nanoparticle-adsorbed carbon nanotube yarns for efficient fiber solar cells. ACS Nano 6(8):7191–7198
97. Cai ZB, Li L, Ren J, Qiu LB, Lin HJ, Peng HS (2013) Flexible, weavable and efficient microsupercapacitor wires based on polyaniline composite fibers incorporated with aligned carbon nanotubes. J Mater Chem A 1(2):258–261

98. Sun X, Sun H, Li H, Peng H (2013) Developing polymer composite materials: carbon nanotubes or graphene? Adv Mater 25(37):5153–5176
99. Weiss NO, Zhou H, Liao L, Liu Y, Jiang S, Huang Y, Duan X (2012) Graphene: an emerging electronic material. Adv Mater 24(43):5782–5825
100. Park S, Ruoff RS (2009) Chemical methods for the production of graphenes. Nat Nanotechnol 4(4):217–224
101. Yang ZB, Sun H, Chen T, Qiu LB, Luo YF, Peng HS (2013) Photovoltaic wire derived from a graphene composite fiber achieving an 8.45 % energy conversion efficiency. Angew Chem Int Ed 52(29):7545–7548
102. Xu Z, Gao C (2011) Graphene chiral liquid crystals and macroscopic assembled fibres. Nat Commun 2:571
103. Xiang CS, Behabtu N, Liu YD, Chae HG, Young CC, Genorio B, Tsentalovich DE, Zhang CG, Kosynkin DV, Lomeda JR, Hwang CC, Kumar S, Pasquali M, Tour JM (2013) Graphene nanoribbons as an advanced precursor for making carbon fiber. ACS Nano 7(2):1628–1637
104. Cruz-Silva R, Morelos-Gomez A, Kim HI, Jang HK, Tristan F, Vega-Diaz S, Rajukumar LP, Elias AL, Perea-Lopez N, Suhr J, Endo M, Terrones M (2014) Super-stretchable graphene oxide macroscopic fibers with outstanding knotability fabricated by dry film scrolling. ACS Nano 8(6):5959–5967
105. Dong ZL, Jiang CC, Cheng HH, Zhao Y, Shi GQ, Jiang L, Qu LT (2012) Facile fabrication of light, flexible and multifunctional graphene fibers. Adv Mater 24(14):1856–1861
106. Carretero-Gonzalez J, Castillo-Martinez E, Dias-Lima M, Acik M, Rogers DM, Sovich J, Haines CS, Lepro X, Kozlov M, Zhakidov A, Chabal Y, Baughman RH (2012) Oriented graphene nanoribbon yarn and sheet from aligned multi-walled carbon nanotube sheets. Adv Mater 24(42):5695–5701
107. Xiang CS, Young CC, Wang X, Yan Z, Hwang CC, Cerioti G, Lin J, Kono J, Pasquali M, Tour JM (2013) Large flake graphene oxide fibers with unconventional 100 % knot efficiency and highly aligned small flake graphene oxide fibers. Adv Mater 25(33):4592–4597
108. Xu Z, Sun HY, Zhao XL, Gao C (2013) Ultrastrong fibers assembled from giant graphene oxide sheets. Adv Mater 25(2):188–193
109. Cong HP, Ren XC, Wang P, Yu SH (2012) Wet-spinning assembly of continuous, neat, and macroscopic graphene fibers. Sci Rep 2:613
110. Xu Z, Liu Z, Sun HY, Gao C (2013) Highly electrically conductive Ag-doped graphene fibers as stretchable conductors. Adv Mater 25(23):3249–3253
111. Shin MK, Lee B, Kim SH, Lee JA, Spinks GM, Gambhir S, Wallace GG, Kozlov ME, Baughman RH, Kim SJ (2012) Synergistic toughening of composite fibres by self-alignment of reduced graphene oxide and carbon nanotubes. Nat Commun 3:650
112. Frank E, Steudle LM, Ingildeev D, Sporl JM, Buchmeiser MR (2014) Carbon fibers: precursor systems, processing, structure, and properties. Angew Chem Int Ed 53(21):5262–5298
113. Hou SC, Cai X, Wu HW, Lv ZB, Wang D, Fu YP, Zou DC (2012) Flexible, metal-free composite counter electrodes for efficient fiber-shaped dye-sensitized solar cells. J Power Sources 215:164–169
114. Zhang F, Niu SM, Guo WX, Zhu G, Liu Y, Zhang XL, Wang ZL (2013) Piezo-phototronic Effect Enhanced Visible/UV Photodetector of a Carbon-Fiber/ZnO-CdS Double-Shell Microwire. ACS Nano 7(5):4537–4544
115. Cai X, Hou SC, Wu HW, Lv ZB, Fu YP, Wang D, Zhang C, Kafafy H, Chu ZZ, Zou DC (2012) All-carbon electrode-based fiber-shaped dye-sensitized solar cells. Phys Chem Chem Phys 14(1):125–130
116. Hamedi M, Forchheimer R, Inganas O (2007) Towards woven logic from organic electronic fibres. Nat Mater 6(5):357–362
117. Hou SC, Lv ZB, Wu HW, Cai X, Chu ZZ, Yiliguma ZDC (2012) Flexible conductive threads for wearable dye-sensitized solar cells. J Mater Chem 22(14):6549–6552

118. Kang TJ, Choi A, Kim DH, Jin K, Seo DK, Jeong DH, Hong SH, Park YW, Kim YH (2011) Electromechanical properties of CNT-coated cotton yarn for electronic textile applications. Smart Mater Struct 20(1):015004
119. Sun GZ, Zheng LX, An J, Pan YZ, Zhou JY, Zhan ZY, Pang JHL, Chua CK, Leong KF, Li L (2013) Clothing polymer fibers with well-aligned and high-aspect ratio carbon nanotubes. Nanoscale 5(7):2870–2874
120. Yang ZB, Deng J, Chen XL, Ren J, Peng HS (2013) A highly stretchable, fiber-shaped supercapacitor. Angew Chem Int Ed 52(50):13453–13457
121. Ma R, Lee J, Choi D, Moon H, Baik S (2014) Knitted fabrics made from highly conductive stretchable fibers. Nano Lett 14(4):1944–1951

Chapter 3
Fiber-Shaped Dye-Sensitized Solar Cell

Abstract This chapter focuses on a burgeoning type of dye-sensitized solar cells (DSCs) with one-dimensional configuration. Distinguished from the conventional planar DSC, the fiber-shaped DSC exhibits a unique structure based on the use of fiber electrode. The working mechanism of DSC is discussed in the beginning. Then, we discuss the fabrication and performance of the two typical architectures in fiber-shaped DSC, i.e., twisted and coaxial DSCs. Next, the fiber-shaped DSC integrated with other functions is presented. Finally, the future development of fiber-shaped DSCs is summarized.

3.1 Overview of Dye-Sensitized Solar Cell

Motivated by photography, Becquerel in 1839 first conducted and reported his experiments on photovoltaic effect. A metal halide-contained solution was illuminated to produce a current between two platinum electrodes. The glory of first sensitization of a semiconductor was bestowed to Vogel, who investigated the sensitization of silver halide emulsions with dyes since 1873. The photoresponse was hence extended to red spectrum promising a panchromatic broad-spectrum photography. The photovoltaic effect was significantly enhanced by using dye sensitizer, and a strengthened photoelectric current was observed by Moser and Rigollot in the 1890s. The photography and photovoltaic effect, developed in parallel in the following centuries, both are virtually recognized as a charge transfer process. Till the twentieth century, it is recognized that effective dye sensitization can be realized providing that the dye molecule was adsorbed on the surface of semiconductor in a closely packed monolayer [1].

The early exploitation of dye sensitization and photoelectrochemistry paves the road towards application. Emerged in the 1980s, the dye-sensitized fractal-type titanium dioxide (TiO_2) was employed as a photoelectrode [2]. However, suffering from a poor light harvesting, the energy conversion efficiency of photoelectrochemical cell was wandering below the barrier of 1 % in a long time, remote to the practical application still. The renaissance of dye-sensitized semiconductors was triggered in 1991, when Prof. Grätzel and coworkers developed a new generation of

H. Peng, *Fiber-Shaped Energy Harvesting and Storage Devices*, Nanostructure Science and Technology, DOI 10.1007/978-3-662-45744-3_3

solar cells applying the newly developed dye-sensitized nanocrystalline TiO_2 that is endowed with a thousand-fold increased surface area beneficial for dye absorption. Thus, an unprecedented energy conversion efficiency over 7 % was achieved [3].

The configuration of planar-shaped DSC inherits the typical sandwich structure developed by Grätzel and coworkers. Two electrodes, a working electrode and a counter electrode, are placed face to face and separated by an electrolyte. The transparent glass with a thin transparent conductive layer of fluorine-doped tin dioxide (FTO) or indium-doped tin oxide (ITO) on one side is employed as the substrate for depositing a thin film of mesoporous semiconductor oxide. The counter electrode is deposited with a catalytic layer including platinum, carbon materials, etc. After the working electrode is immersed into dye for absorption, the two electrodes are sealed followed by injecting the electrolyte.

Based on the conventional planar structure, a DSC can be also made into a one-dimensional configuration. A wide range of fiber materials including carbon fibers, metal fibers, and polymer fibers are adopted to construct the fiber-shaped DSCs. The configuration of fiber-shaped DSCs can be classified into two types, i.e., twisted and coaxial structures, which will be discussed in detail later.

A panorama view of the working process in the DSC is summarized below. When illuminated, the dye molecules harvest photons and get excited. The excited dye molecules then inject electrons into the conduction band of the nearby semiconductor. Electrons transport through the semiconductor layer and travel throughout the external circuit. The triiodide ions in electrolyte obtain electrons at the counter electrode and are reduced into iodide ions which release electrons to the oxidized dye molecules to regenerate the dyes. Meanwhile, the charge injection and transportation are accompanied by the recombination, which diminishes the output performances of the DSC. In the next sections, the mechanism is detailed by energetic and kinetic views.

3.1.1 Principle in an Energetic View

To clarify the working mechanism of DSCs, an energetic view of the involved materials including dye, semiconductor, and redox couple is presented in Fig. 3.1 [4]. The relative position of those energy levels is crucial to the performance of a DSC. Several points should be noted and cleared to prevent any confusion. First, the ordinate in Fig. 3.1 represents the internal energy (not the free energy), because a significant contribution from the configuration entropy, which arises from the huge energy state accessible to the charge carrier, is not concerned. Second, the energy levels of the dye molecules are conventionally presented by the energy of the highest occupied molecular orbital (HOMO) and lowest unoccupied molecular orbital (LUMO), which are usually obtained from calculation involving several approximations. Third, the electrochemical potential of electrons in a semiconductor is normally identified as the Fermi level (E_F) with dimension of energy, while that for a redox couple in electrolyte is normally referred as the redox potential

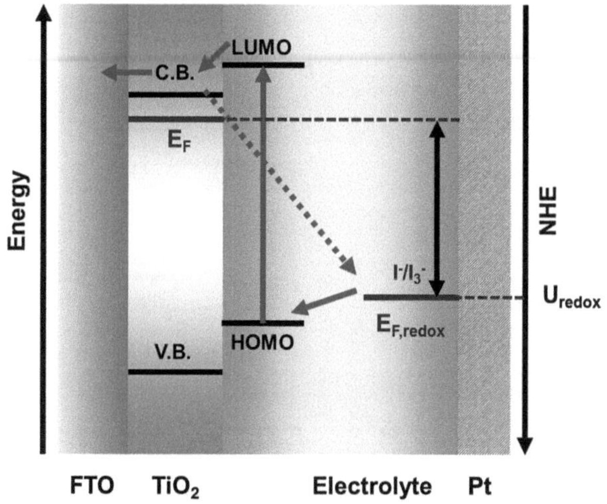

Fig. 3.1 Schematic illustration of the working mechanism of a DSC

(U_{redox}) with dimension of voltage. An empirical relationship is given below which relates the Fermi level ($E_{F,\text{redox}}$) of the redox couple and its redox potential. The Fermi levels of a semiconductor and electrolyte equalize at equilibrium [5].

$$E_{F,\text{redox}}[eV] = -(4.6 \pm 0.1) - eU_{\text{redox}}[V] \tag{3.1}$$

A special attention should be paid on the different energy scales applied to depict the electrochemical potential of a semiconductor and redox couple, as presented in Fig. 3.1. We prefer to use the vacuum energy as reference to depict the energy level of a semiconductor and normal hydrogen electrode (NHE) as reference for a redox couple. With the help of the above formula, the standard potential of a redox couple can be readily transformed to the electrochemical potential.

3.1.1.1 Energy Level in Semiconductor

The energy level of semiconductor will be affected upon contacting the electrolyte. In absence of redox couple, the solvation effect of the solvent together with Helmholtz layer dipole arising from dye absorption will shift the conduction band of semiconductor towards the vacuum level. When redox couple is added, the electron will transfer from the semiconductor to the hole carrier in the electrolyte, driven by the energy differences of Fermi level and redox potential. Until $E_F = E_{F,\text{redox}}$, the semiconductor is depleted. However, because the particle size is much smaller than the Debye length, no space charge layer is formed inside individual particles and the band bending is negligibly small. In the dark, the Fermi level of the semiconductor is equal to the redox potential of the electrolyte, and when illuminated and followed by electron injection, the quasi-Fermi level of the

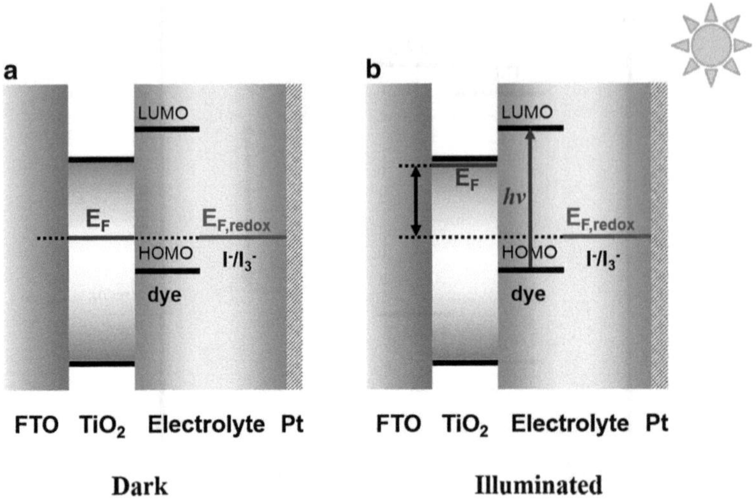

Fig. 3.2 Schematic diagram of energy level of a semiconductor contacting with electrolyte before (**a**) and after illumination (**b**)

semiconductor is elevated close to the conduction band edge, as displayed in Fig. 3.2. Therefore, the produced photovoltage originated from the energy difference between the electron Fermi level in semiconductor and the redox potential of the electrolyte [6].

Note that the conduction band edge (E_c) of semiconductor can be tuned in a wide range in the presence of cations. For example, 4-*tert*-butylpyridine (TBP), widely adopted as an additive in the electrolyte, can significantly improve the photovoltage and conversion efficiency [5].

3.1.1.2 Energy Level of Redox Couple

Generally, the redox couple in electrolyte is iodide/triiodide (I^-/I_3^-). The ingredient of the electrolyte contains 0.1 M lithium iodide (LiI) and 0.05 M iodine (I_2). The triiodide ions are produced by the composition of iodine and iodide:

$$I_2 + I^- = I_3^-$$

The redox potential of the I^-/I_3^- is given by the Nernst equation:

$$I_3^- + 2e^- = 3I^-$$

$$E_{F,\mathrm{redox}} = E^{0'}_{F,\mathrm{redox}} - k_B T \ln\left(\frac{[I_3^-]}{[I^-]^3}\right) \tag{3.2}$$

where $E^{0'}_{F,\mathrm{redox}}$ is the formal redox potential, k_B is the Boltzmann constant, and T is the temperature. Apart from the above redox reaction, several one-electron reactions can take place as well involving the iodine radicals (I•) and diiodide (I_2^{-}•), as shown below:

$$2\mathrm{I}\bullet = \mathrm{I}_2$$

$$\mathrm{I}\bullet + \mathrm{I}^- = \mathrm{I}_2^{-\bullet}$$

The standard potential of iodide/triiodide redox couple is 0.35 V vs. NHE, while the oxidation potential of the dye is normally 1.1 V vs. NHE. The difference of the energy levels provides the driving force for the dye regeneration [5, 7].

3.1.1.3 Energy Level of Excited Dye Molecule

As illustrated in Fig. 3.3, the reduction of dye molecules takes place provided that the electron transfers from the donor to the unoccupied level of the molecule. Likewise, the oxidation of dye molecules occurs by the electron transfer from the lower occupied state to an acceptor molecule. The Fermi levels of the two redox systems are $E^{0}_{F,\mathrm{redox}}$ (D^-/D) and $E^{0}_{F,\mathrm{redox}}(D/D^+)$. The difference of them can be roughly described as the energy difference of LUMO and HOMO in dye molecule. Figure 3.3 draws a clear picture that the electron transfer is favorable from a higher level to an acceptor or from a donor to a half-filled orbital. An excited dye molecule is easier for oxidation or reduction since the excitation energy is stored in the molecule. For the classical dye molecule, $RuL_2(NCS)_2$, the $E^{0}_{F,\mathrm{redox}}(D/D^+) = -5.6$ eV, and after excitation, $E^{*}_{F,\mathrm{redox}}(D/D^+) = -3.85$ eV [8].

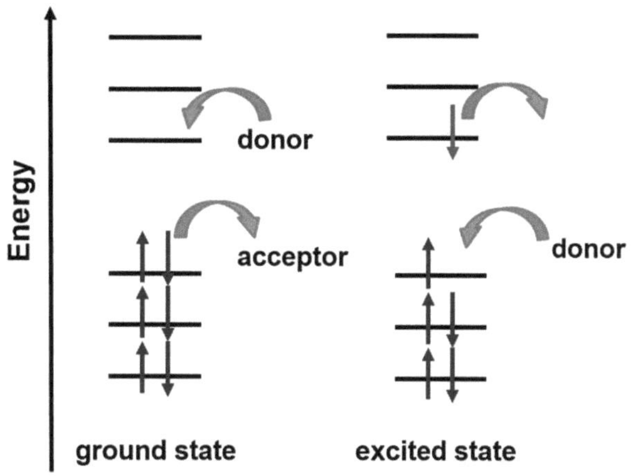

Fig. 3.3 Molecular energy levels in the ground and excited states

3.1.2 Principle in a Kinetic View

3.1.2.1 Charge Separation and Injection

The breakthrough in 1991 was benefited from the preparation of nanocrystalline TiO_2, whose particle size is much smaller than its Debye length thus all the particles are depleted upon contact with the electrolyte. As a result, the produced charges cannot be separated by the electric field in the space charge layer. In fact, the charge separation in DSC is driven by the gradient of free energy or Fermi levels of electrons and holes.

The photoinduced charge transfer from a semiconductor to adsorbate species is determined by the band energy positions of the semiconductor and the redox potentials of the adsorbate species. The relative energy level of the semiconductor and adsorbate species provides the thermodynamic impetus of the charge transfer. The energy difference of LUMO and conduction band provides an enthalpic driving force for electron injection, and the difference of redox potential and HOMO energy presents a driving force for hole injection.

The charge separation of the photoexcited electron and hole occurs by the ultrafast injection from the excited dye to the semiconductor oxide. As for N719, light absorption is of metal-to-ligand charge transfer character. The excitation process promotes electrons from the metal center to the carboxylate bipyridyl ligand anchoring on the surface of a TiO_2 particle (Fig. 3.4). Generally, the excited-state charge separation takes place from the π* orbitals of the ligands to the conduction band of TiO_2. Femtosecond transient absorption spectroscopy is used to determine the kinetics of the injection process. Although the detailed mechanisms are vigorously debated, the time scale of injection process can be confirmed within 10^{-12}–10^{-13} s [4, 5, 9].

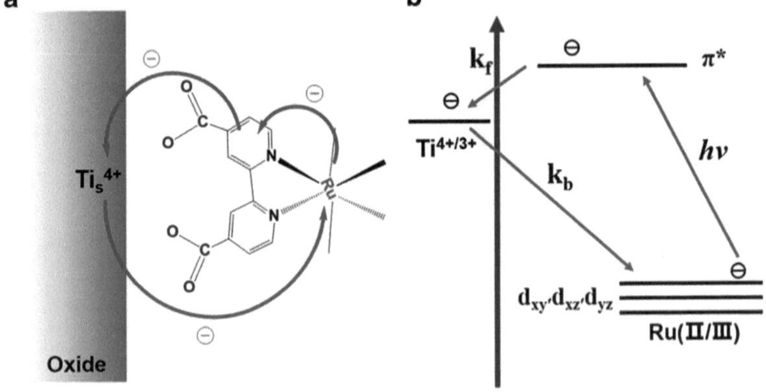

Fig. 3.4 Schematic illustration of metal-to-ligand charge transfer. **a** A ruthenium complex anchored on the surface of a titanium dioxide particle. **b** The corresponding molecular orbital diagram of **a**

The electron injection efficiency is defined as the fraction of the photons captured by the dye that are converted into conduction band electrons:

$$\varphi_{\text{inj}} = \frac{k_{\text{inj}}}{k_{\text{inj}} + k_d} \tag{3.3}$$

where k_{inj} and k_d are the rate constants for electron injection and decay of the excited dye, respectively. Benefited from the ultrafast charge injection (~160 fs), the DSC can deliver a high electron injection efficiency of more than 90 %.

The injection process is concomitantly incorporated with a charge recombination which occurred between the electrons in the conduction band and the oxidized dye. The kinetic parameter of the recombination process is k_b, which is several orders of magnitude smaller than k_{inj}. Moreover, the π^* orbitals of the carboxylated bipyridyl ligand overlap favorably with the conduction band of the titanium dioxide, rather than the *d* orbitals localized on the oxidized dyes. Thus, the charge recombination between the conduction band and the oxidized dyes is negligible.

3.1.2.2 Charge Recombination

The injected electrons will travel a long path from the semiconductor to the current collector. During this distance, the generated photocurrent will be attenuated by a recombination process. Three species can scavenge the electrons: the oxidized dye molecules, oxidized electrolyte species, and contaminants in the system. As clarified previously, the electron transfer towards the oxidized dye molecules is energetically and kinetically unfavorable. And the electron transfer to the oxidized dye is always competing with the dye regeneration, i.e., the oxidized dye is reduced by the iodide. Regarding the huge number of iodide ions in the electrolyte, the electron transfer to the oxidized dye is negligible. Therefore, the recombination principally occurs between electrons injected to TiO_2 and triiodide at the electrolyte/TiO_2 interface.

$$2e^- + I_3^- = 3I^-$$

The kinetics of the recombination is still under debate, as reviewed by Pettersson H et al. [5]. As expected, the recombination rate is increased with increasing triiodide concentration and the reaction is temperature dependent. It can be suppressed by additives in the electrolyte including 4-*tert*-butylpyridine. The recombination is reflected in the fill factor of the resulted current–voltage curve. The fill factor is diminished when the electron transfer from TiO_2 to I_3^- rises with increasing photovoltage [6].

3.1.2.3 Open-Circuit Photovoltage

When a DSC is illuminated, a high photovoltage is expected. In the dark, the Fermi level of the system is determined by the redox potential of the electrolyte. Upon illumination and subsequent electron injection from the dye, the electron Fermi level in TiO_2 shifts towards the conduction band edge while the counter electrode remains its redox potential.

After a dye molecule is excited, the electrons are injected to the conducting band of the TiO_2. The injection rate is determined by the light intensity and the surface coverage of adsorbed dye. Thus, the injection rate can be expressed as

$$\upsilon_{\text{inj}}(\lambda, x) = \eta_{\text{inj}}\alpha(\lambda)I_0 e^{-\alpha x} \tag{3.4}$$

where η_{inj} is the efficiency of electron injection from the excited dye and $\alpha(\lambda)$ is the absorption coefficient that is related to the molar concentration and molar absorption coefficient of the dye.

The open-circuit voltage is compromised by the recombination, whose rate is determined as

$$\upsilon_{\text{rec}} = k_{\text{I}_3^-} n \left[\text{I}_3^-\right] \tag{3.5}$$

For simplification, we assume that the electron transfer to oxidized dyes through the substrate is negligible. Then, the electron density in steady state is given by

$$n = \frac{\upsilon_{\text{inj}}}{k_{\text{I}_3^-}\left[\text{I}_3^-\right]} = \upsilon_{\text{inj}}\tau_n \tag{3.6}$$

where the τ_n is the electron lifetime. Provided that the electrons are free to move in the conduction band of the semiconductor, according to the Fermi–Dirac function, the density of electrons in the conduction band is determined as

$$n_c = N_c \exp\left(-\frac{E_{\text{CB}} - E_F}{k_B T}\right) \tag{3.7}$$

The Fermi level in the dark is equal to the redox Fermi level; thus, the electron density is

$$n_c^0 = N_c \exp\left(-\frac{E_{\text{CB}} - E_{F,\text{redox}}}{k_B T}\right) \tag{3.8}$$

Thus, the photovoltage is given by the energy difference between the electron Fermi level of TiO_2 and the redox potential of the electrolyte ($E_{F,\text{redox}}$).

$$\begin{aligned} \text{q}V_{\text{OC}} &= E_F - E_{F,\text{redox}} \\ &= (E_{\text{CB}} - E_{F,\text{redox}}) - (E_{\text{CB}} - E_F) = k_B T \ln\frac{n_c}{n_c^0} \end{aligned} \tag{3.9}$$

Equation 3.6 reveals that to achieve an open-circuit voltage of 0.77 V, the ratio of electron densities of the conduction band n_c/n_c^0 should be 10^{13}. Provided that $E_F - E_{F,\text{redox}} = 1\ \text{eV} N_c = 10^{21}\ \text{cm}^{-3}$, the electron density in the dark is $10^4\ \text{cm}^{-3}$, and a photovoltage of 0.77 V corresponds to a conduction band electron density of 10^{17} cm^{-3} [7].

From the kinetic view, the expression of photovoltage can also be deduced from the *p-n* junction solar cell:

$$V_{OC} = \frac{kT}{e} \ln\left(\frac{I_{inj}}{I_0} + 1\right) \tag{3.10}$$

where k is the Boltzmann constant, T is the temperature, and I_0 is the dark current arising from recombination.

$$I_{inj} = q\eta\Phi_0 \tag{3.11}$$

$$I_0 = qn_0k_{rec}\left[I_3^-\right] \tag{3.12}$$

where η is the quantum yield for photogenerated electrons, Φ_0 is the incident photon flux, n_0 is the equilibrium electron density of the conduction band in the dark, and k_{rec} is the bimolecular rate constant for recombination. Substituting (3.12) and (3.11) into (3.10), we obtain

$$V_{OC} = \frac{kT}{q} \ln\left(\frac{\eta\Phi_0}{n_0k_{rec}\left[I_3^-\right]} + 1\right) \tag{3.13}$$

Considering $I_{inj} \gg I_0$, then the expression can be simplified as

$$V_{OC} = \frac{kT}{q} \ln\left(\frac{\eta\Phi_0}{n_0k_{rec}\left[I_3^-\right]}\right) \tag{3.14}$$

This expression reveals that the performances of DSCs are less sensitive to illumination intensity compared with traditional *p-n* junction solar cells. The photovoltage is predicted to be increased by 59 mV in every tenfold increase of illumination intensity [10].

3.1.3 Materials

3.1.3.1 Semiconductor

TiO_2 is popular in DSC because it is photostable and nontoxic and has a high refractive index (n=2.4–2.5). TiO_2 has three crystal forms in nature: rutile, anatase, and brookite. Anatase is widely used due to a bandgap of 3.2 eV that is higher than that of rutile (3.0 eV) and has a higher conduction band edge energy, leading to a higher Fermi level and photovoltage. Many a structure and morphology of TiO_2 has been developed in recent years, spanning from nanoparticles, nanotubes, and nanowires to nanosheets. The conventional method for synthesizing the TiO_2 nanoparticles is involved in hydrolyzing a titanium precursor like titanium alkoxide, followed by hydrothermal growth and crystallization. The resultant particles are mixed with polymers to produce slurry and then deposited on the conducting glass using doctor blade or screen print techniques. Afterwards, the film is sintered at 450 °C [5]. The thickness of the yielded film is controlled at ~10 μm and the diameter of TiO_2 nanoparticles is ~20 nm. Then, the film is further treated with titanium tetrachloride, coated with an ultrathin TiO_2 shell (~1 nm) on the mesoporous nanoparticles followed by heating treatment. The treatment increases the roughness, thus enhancing the dye absorption and improves the injection efficiency [11].

The TiO_2 nanoparticles are used for the semiconductor layer in the early attempts of fiber-shaped DSCs. Nowadays, vertically aligned titanium dioxide nanotubes, prepared by anodization, are preferred in fiber-shaped DSCs. The morphologies of the resulting nanotubes, length, diameter, and wall thickness can be tuned by anodization potential, time, temperature, and electrolyte during the synthesis, respectively [12]. Titanium dioxide nanotubes as working electrode exhibit higher collection efficiencies for photoinjected electrons [13–15].

Apart from titanium dioxide, zinc oxide (ZnO) is another semiconductor employed in DSC. The bandgap and conduction band edge of the zinc oxide is near that of titanium dioxide, and the zinc oxide has a higher electron mobility. But compared with titanium dioxide, the zinc oxide has a relatively poorer chemical stability and is unsustainable to acid and base [5, 16].

3.1.3.2 Dye

Historically, titanium dioxide is popular for photocatalyzed water cleavage, following the work of Fujishima and Honda in 1972 [17]. It was latterly developed into photoelectrochemical cells. However, the future of photoelectrochemical cell is clouded by a tricky issue that titanium dioxide is incapable of using visible light because of its wide bandgap, while other semiconductors with smaller bandgaps are insufferable for photocorrosion [18].

This dilemma was resolved by separating the light absorption and charge transport functions, which was integrated in the case of traditional TiO_2-based photocatalytic process. A photosensitizer is introduced to adsorb the sunlight and inject the excited charges to a wide-gap and photostable semiconductor. Thus, benefited from the photosensitizers, the charge separation can be realized with incident light with energy less than bandgap. That is what "sensitize" means in DSC.

A large variety of dyes are used as photosensitizers. Among the intensively developed dyes, *cis*-bis(thiocyanato)bis(2,2′-bipyridyl-4,4′-dicarboxylato)ruthenium(II), known as N719, is the most successful and widely used charge transfer sensitizer. Figure 3.5 displays the chemical structure and its incident photon-to-

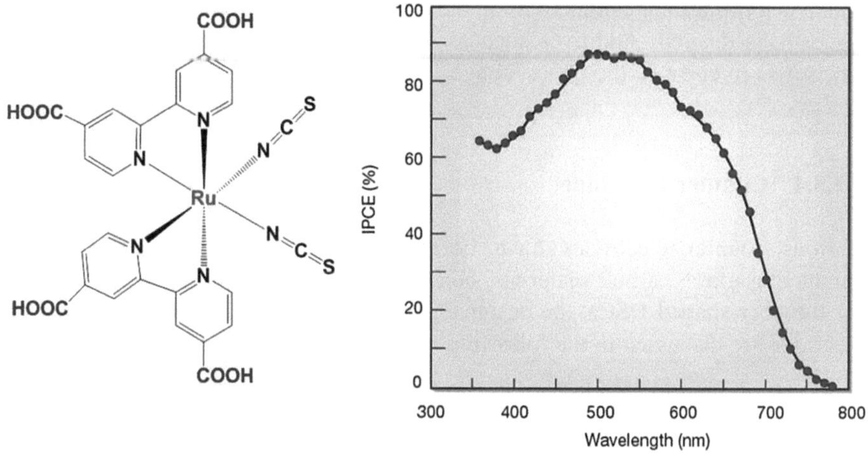

Fig. 3.5 Chemical structure of N719 (*left*) and its incident photon-to-current conversion efficiency curve (*right*) (Reprinted by permission from Nature Publishing Group: Ref. [18], copyright 2001)

current conversion efficiency (IPCE) curve [18]. N719 molecules are endowed with several characteristics making them a successful photosensitizer. Structurally, the N719 molecule has four carboxylate groups (−COOH) as anchoring groups, which enable the dye molecule to firmly bind onto the semiconductor surface. The IPCE curve reveals that the absorption spectrum of N719 covers the visible region and even part of near-infrared region. Energetically, the energy levels of N719 are well matched with those of semiconductor and redox couple in the electrolyte. As shown in Fig. 3.1, the excited level in energy of the dye is above the conduction band edge of the semiconductor, so that the excited electrons can be injected to the semiconductor. The HOMO of the dye is lower than the redox potential of the electrolyte, which guarantees its regeneration. The N719 is also reported for a superb photostability. It can sustain more than 5×10^7 redox cycles without an observable loss of performance [3, 4]. Presently, N719 is mainly adopted as the photosensitizer for the fiber-shaped DSCs still. In our following discussion, the DSC is fabricated based on the dye of N719 if not specified.

3.1.3.3 Redox Electrolyte

I^-/I_3^- is the most favorable redox couple adopted in DSCs. Generally, the electrolyte contains mixtures of iodides such as LiI, NaI, KI, and I_2 dissolved in nonprotonic solvents such as acetonitrile. The performances of DSCs depend on counter cations of iodides arising from different ion conductivities in the electrolyte or adsorption on the semiconductor surface, leading to a shift of the conduction band level. The performances of DSCs are also affected by the viscosity of solvents, which determines ion conductivity in the electrolyte. Therefore, a solvent with low viscosity is desired. Additives such as 4-*tert*-butylpyridine are contained in the electrolyte to enhance the photovoltage.

3.1.3.4 Counter Electrode

Various counter electrodes have been developed spanning from platinized conducting glass, carbon materials, conducting polymers, and cobalt sulfide. As for the fiber-shaped DSCs, the designation and performances of different counter electrode are discussed in the following sections.

3.1.4 Characterization

The standard efficiency measurements of DSCs are conducted under AM 1.5 G solar radiation spectrum, which is normalized so that the integrated irradiance (the amount of radiant energy received from the sun per unit area per unit time) is 1,000 W m^{-2}.

The energy conversion efficiency of a DSC is given as

$$\eta = \frac{V_{OC} J_{SC} FF}{I_0} \tag{3.15}$$

where V_{OC} is an open-circuit voltage; J_{SC} is short-circuit current density; I_0 is the intensity of the incident light, which is 1,000 W m^{-2} at AM 1.5 G; and FF is fill factor, which is defined as the ratio of the maximum power (P_{max}) and the product of V_{OC} and J_{SC}:

$$FF = \frac{P_{max}}{V_{OC} J_{SC}} \times 100\ \% \tag{3.16}$$

The maximal power is obtained at the point of *J–V* curve where the product of the photocurrent and photovoltage is maximal.

Another important parameter to characterize the DSC is the incident photon-to-current conversion efficiency (IPCE), which corresponds to the photocurrent density produced in the external circuit under monochromatic illumination of the cell divided by the photon flux that strikes the cell and provides practical information about the monochromatic quantum efficiencies of a solar cell [5].

$$\text{IPCE} = \frac{J_{SC}(\lambda)}{e\Phi(\lambda)} = 1,240 \frac{J_{SC}(\lambda)[\text{A cm}^{-2}]}{\lambda[\text{nm}] I_0(\lambda)[\text{W cm}^{-2}]} \tag{3.17}$$

3.1.5 Summary

DSC is considered as a technology between the second- and third-generation solar cells. Compared with the conventional photovoltaic devices, the DSC displays the following merits:

1. High energy conversion efficiency: currently, dye-sensitized solar cells deliver a high conversion efficiency over 10 %, which is comparable to the silicon-based solar cells.
2. Low cost: the fabrication of DSC costs less because the fabrication process is exempt from the sophisticated technologies involved in fabricating silicon-based solar cells. Semiconductors such as TiO_2, dye, and iodine are abundantly available.
3. Diversity: their configurations provide more possibilities for designing transparent, colorful devices for consumer products. More importantly, the glass substrate can be substituted with flexible, stretchable substrate to make the DSC flexible, deformable, and light weight, which is favorable for the wearable, flexible, and portable devices.

3.2 Overview of Fiber-Shaped Dye-Sensitized Solar Cell

The conception of fiber-shaped DSCs, or as they called "solar cells in a fiber form," can be traced as early as 2001 [19]. The proposed structure of the solar cell was not far from the currently adopted configuration. Quite similarly, instead, it displays a coaxial structure: a stainless steel was covered with a layer of dye-absorbed TiO_2 particles and a transparent conductive polymer, as counter electrode, was wrapped outside. The two electrodes were sealed in a transparent tube followed by injecting electrolyte. The as-fabricated solar cell, 10 cm in length, was able to deliver a voltage of 0.3–0.35 V. However, suffering from the low conductivity of the polymer electrode as well as the corrosive electrolyte, only several microamperes of current was detected. The premiere of the fiber-shaped solar cell may not be recognized as a success, but not a failure either. The proposed conception of "solar cells in a fiber form" laid the foundation of the fiber-shaped DSC and paved the first stepping-stone towards its application pertaining to wearable electronic devices. More importantly, this visionary work has inspired the scientists that the probable approach towards high performance is to develop the competent fiber electrode.

Ahead of the fiber-shaped DSC, polymer solar cells and organic solar cells were successfully fabricated into a fiber format and achieved a conversion efficiency of 0.6 % and 0.5 %, respectively [20, 21]. Until 2008, echoing the preceding inventions, the first efficiency-recorded fiber-shaped DSC appeared [22]. Not too long ago did the fiber-shaped DSCs emerge, though they have experienced a rapid growth in recent years. Inheriting the merit of one-dimensional configuration, the fiber-shaped DSCs are endowed with small size, light weight, and flexibility. They can harvest light all around and take advantage of the diffuse light from the surroundings. The fiber-shaped DSC is always progressing in parallel with the development of fiber electrode. In this section, we will take a general view of the emergence and evolution of fiber-shaped DSCs classified into several categories [23].

3.2.1 Metal-Based Electrode

The widely used metal for counter electrode is platinum for its high conductivity and superior catalytic ability. Considering the high expense of platinum, an alternate electrode was developed by electrochemical depositing Pt particles on a metal wire. The working electrode is delicately designed by compositing TiO_2 nanomaterials onto a metal wire, such as Ti wire and stainless steel wire. Various methods have been developed including coating TiO_2 nanoparticles and modifying with TiO_2 nanowires and nanotubes.

Initially, the fiber-shaped DSC was materialized by twisting two metal electrodes, stainless wire coated with dye-sensitized TiO_2 as the working electrode and platinum wire as the counter electrode [22]. A conventional liquid electrolyte was applied. The as-fabricated DSC displays a typical one-dimensional configuration with two fiber electrodes symmetrically twisted. Though closely contact, the surface-coated TiO_2 as a barrier layer prevented the short circuit between the two electrodes. One of the advantages proclaimed of the fiber-shaped DSC is that their external surfaces are available for light from arbitrary angle. Thus, their photocurrents are immune from the incident angle of the light. The screw pitch and thickness of the TiO_2 layer play a critical role in the performances of the DSCs. The major concern is the compromise between the photocurrent generation and the back transfer. After the optimization, several centimeters of fiber-shaped DSC obtained an open-circuit voltage of 0.61 V, short-circuit current of 0.06 mA, and fill factor of 0.38, and the final energy conversion efficiency was recorded 0.27 %.

The stainless was afterwards replaced by Ti wire, and porous TiO_2 nanowire was grown on the surface of the Ti wire through hydrothermal process. An improved efficiency of 0.86 % was achieved [24]. During early times, the efficiencies of fiber-shaped DSCs are wandering below 1 %, which are dwarfed by their planar-shaped analogues. A breakthrough was made in 2010 [25]. A Ti wire vertically grown with TiO_2 nanotube arrays was employed as the working electrode. The aligned TiO_2 nanotubes were favorable for the electron transfer with smaller resistance and suppressed recombination. Facing in parallel with the Pt counter electrode, the fabricated solar cell provided a higher conversion efficiency of 2.54 %, which was mainly attributed to the significant raise of photocurrent density. The recorded short-circuit current density was several times higher than previous results. However, the fill factor, strongly dependent on cell structure and assembling process, decreased as a result of the configuration that the two electrodes were placed in parallel. A targeted amelioration was conducted by intertwining the Pt wire around the TiO_2nanotube-modified Ti wire, and the conversion efficiency increased to 6.72 % [26]. Note that the modified Ti wire was treated with $TiCl_4$ to introduce TiO_2 nanoparticles. The length of the TiO_2 nanotube will affect the performances as the trade-off between the dye absorption and charge recombination.

Another approach to prepare the working electrode is dip-coating TiO_2 layer on the Ti wire. The thickness can be tuned in the range of 10–40 μm. The thicker layer promised a larger amount of dye absorption while increasing the recombination rate. Making the solar cell into a longer scale will sacrifice its conversion efficiency. This method enabled the solar cell to be scaled up to a larger size with limited efficiency decay. An optimized efficiency of 5.41 % was yielded in a 9.5 cm cell [27]. As we declared previously, the fiber-shaped DSC is beneficial for harvesting the diffuse light from surroundings. When a parabolic reflector was used to focus diffuse light, the conversion efficiency was further increased to 7.02 % [28].

Apart from the conventional TiO_2 nanotubes and films, other architectures, such as TiO_2 nanowires and porous network, were also coated as the outer sheath of Ti wire [29, 30]. All the working electrodes based on metal wire exhibited a typical core–sheath structure: the photoactive layer outside as sheath and the metal wire inside as core. The sheath, commonly dye-absorbed TiO_2 nanomaterials, can in situ be generated from the core or, otherwise, introduced as a second phase. From the above discussion, we can deduce that a successful working electrode should afford a large dye absorption and low transfer resistance. The former requires a tremendous surface area, while the latter prefers a short transfer distance and limited interfacial boundaries. They are, however, rarely satisfied concurrently. Therefore, compromises are often made to strike a balance between the dye absorption and charge recombination. Besides, since the fiber-shaped DSC is proposed for applications pertaining to wearable electronic devices, the flexibility should be set as an essential quality to evaluate the electrode. Unfortunately, the mechanical investigation on the Ti wire coated with porous TiO_2 layer did not provide any encouraging results. The TiO_2 layer was readily to crack and damage when bended which is unsustainable for repeated deformation [31, 32].

3.2.2 Metal-Free Electrode

The development of metal-free electrode was motivated by finding a replacement of Pt wire as counter electrode, concerning the high price of the noble metal. Carbon materials, like graphite, carbon black, carbon nanotubes, and graphene, were verified as promising candidates as counter electrodes in planar-shaped DSCs. It is well reasoned to introduce fibrous carbon materials into fiber-shaped DSC to replace the Pt wire. As a creative endeavor, a commercially available carbon fiber was employed as the counter electrode. The carbon fiber, comprising bundles of monofilaments, is flexible and strong and has a high conductivity of 10^2 S cm^{-1} and large surface area of 1.3 m^2 g^{-1}, which make it competent as an electrode. A decent conversion efficiency of 2.7 % was realized using a 230 μm carbon fiber as counter electrode. The performance of carbon fiber can be enhanced by sputter platinum onto its surface. The yielded platinized carbon fiber has an improved catalytic ability and conductivity, as confirmed by the electrochemical impedance spectrum, rendering a conversion efficiency up to 5.1 % [33].

Carbon nanotube (CNT) fiber is another carbon material adopted as counter electrode, which is renowned for its superior conductivity (10^2–10^3 S cm^{-1}) and mechanical strength. A CNT/Pt composite fiber was produced by embracing Pt nanoparticles within the yarns. The Pt nanoparticles were uniformly distributed along the CNT bundles, ensuring an excellent catalytic activity. The conversion efficiency based on the composite fiber was 4.85 %, comparable with a Pt wire electrode [34].

As the fibrous carbon is more flexible and light weight than the metal wire, a bold effort was conducted to fabricate an all-carbon electrode. The TiO_2 can be either coated onto the carbon fiber or in situ grown as nanowires. The counter electrode was prepared by platinizing carbon fiber or coating ink carbon. The as-fabricated solar cells can deliver a conversion efficiency up to 2.48 % [35, 36].

3.3 Fiber-Shaped Dye-Sensitized Solar Cell in a Twisting Structure

Traditional planar DSCs are fabricated in a sandwich configuration. Things are quite different for the fiber-shaped DSC. Up to date, in many cases, fiber-shaped DSCs are made from two fiber electrodes, one as working electrodes modified with photoactive materials and another as counter electrodes to catalyze the redox reaction in the electrolyte [23]. To effectively fabricate the fiber-shaped DSC, fiber working electrodes and counter electrodes should be skillfully prepared. For the fiber electrode, they should be flexible and have high electrical conductivity. Currently, carbon-nanostructure fibers, such as carbon nanotube (CNT) fiber and graphene fiber, and traditional metal wires represent the most studied materials in the fiber-shaped DSC. In addition, in order to enhance energy conversion efficiency and stability of the devices, iodine-free and quasi-solid-state electrolytes are also introduced to the fiber-shaped DSC.

3.3.1 *Working Electrode*

For the traditional planar DSC, the transparent conductive glass (e.g., fluorine-doped tin oxide, FTO) coated with a layer of porous semiconductor metal oxide (e.g., titanium dioxide, TiO_2) is used as the working electrode. However, the planar and rigid working electrode is not suitable for the fiber-shaped DSC. Carbon nanotubes have been introduced to fabricate high-efficiency solar cells due to their extremely high surface area and superior electrical properties. For example, titanium dioxide nanoparticle films modified with carbon nanotubes have been shown to increase the roughness factor and decrease the charge recombination of electron–hole pairs [37]. To really utilize the carbon nanotubes in the fiber-shaped DSC, it is critical to assemble the random carbon nanotube into a macroscopic fiber to maintain their extremely high surface area and superior electrical properties.

3.3.1.1 Carbon Nanotube Fiber

To expand practical applications of carbon nanotubes (CNTs) that are widely studied for remarkable electronic, mechanical, and thermal properties, CNTs have been recently made into continuous fibers, in which the CNTs are highly aligned to extend their excellent properties from nanoscale to macroscopic scale [38]. For instance, these CNT fibers show electrical conductivities of 10^2–10^3 S cm^{-1} and tensile strengths of 10^2–10^3 MPa. They can be continuously made at a large scale from several to tens of micrometers. The fiber diameter can be controlled ranging from micrometers to millimeters by varying the initial width of ribbon. Moreover, they are more flexible than metal wires, carbon fibers, and other available fiber materials such as modified engineering fibers. In addition, the designed nanostructure provides a higher interface area which favors the charge separation and transport. The above unique advantages enable CNT fibers to be an ideal candidate for the fiber-shaped DSC.

As shown in Fig. 3.6a, the CNT fiber is flexible and does not break after being bent, folded, or even tied many times [39].High-resolution transmission electron microscopy shows a multiwalled structure for building nanotubes with diameter of approximately 8.5 nm in the fiber (Fig. 3.6b). The CNT fiber shows a high surface area typically on the order of 10^2 m^2 g^{-1}. Therefore, kinds of materials could be easily physically incorporated or chemically bonded onto CNTs in the fiber. As shown in Fig. 3.6c, after incorporation of 10,12-pentacosadiynoic acid followed by photopolymerization, the fluorescent polydiacetylene is well distributed in the fiber. These features indicate that the CNT fiber could be a promising candidate for fiber working electrode. Figure 3.6d, e compares SEM images of the CNT fiber before and after incorporation of the N719 dye molecules. Raman spectra further confirmed the existing interactions between N719 molecules and CNTs.

The CNT/N719 composite fiber is used as the fiber working electrode to fabricate the DSC [39]. As shown in Fig. 3.7a, the DSC based on the composite fiber (6 mm in diameter and 5 mm in length) produced a short-circuit density (J_{SC}) of 10.3 mA cm^{-2}, open-circuit voltage (V_{OC}) of 0.47 V, and fill factor (FF) of 0.45. The power conversion efficiency (η) can reach 2.2 %. The recorded photocurrent verifies that photocurrent is generated by electron injection from excited N719 molecules into the conduction band of CNTs followed by transport along the CNT fiber, collection by conducting glass, and flow

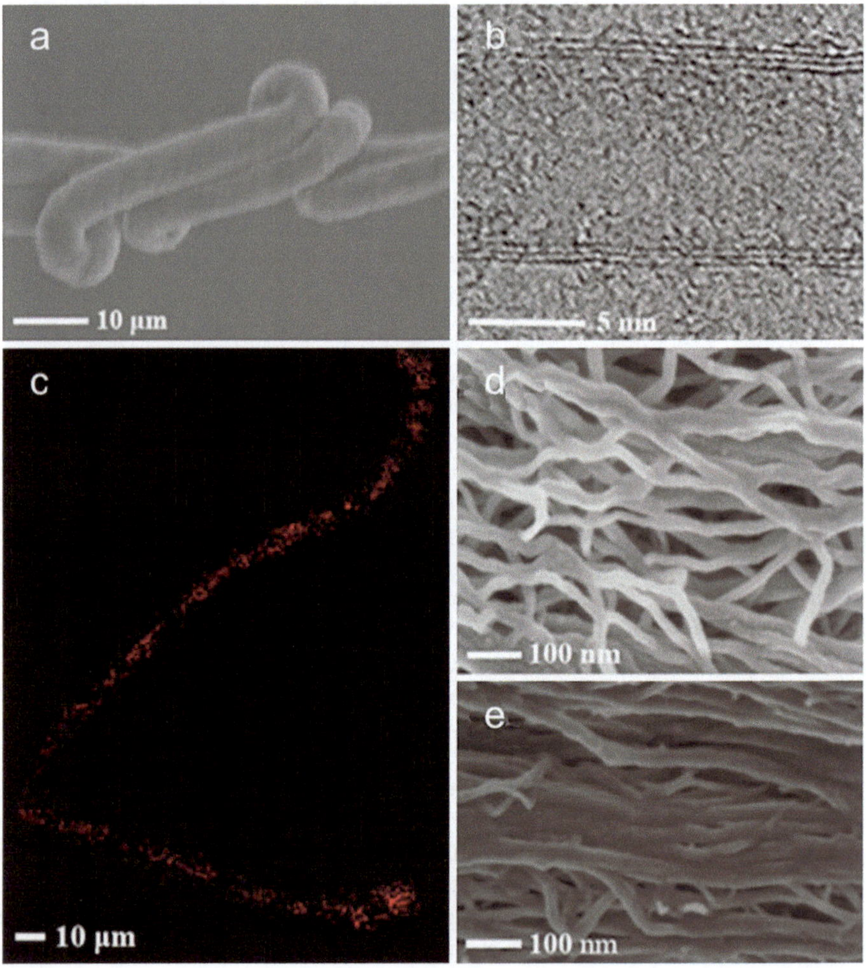

Fig. 3.6 **a** SEM image of a CNT fiber being tied into knot. **b** High-resolution TEM image of a CNT. **c** Confocal laser scanning microscopy image of a CNT fiber after incorporation of diacetylene followed by photopolymerization. **d** and **e** SEM images of a CNT fiber before and after incorporation of N719, respectively (Reproduced from Ref. [39] by permission of John Wiley & Sons Ltd)

through external circuit to the counter electrode (Fig. 3.7b). Although the solar cell is successfully fabricated from the fiber working electrode, it is not a really fiber-shaped DSC due to its use of the planar counter electrode. Moreover, the open-circuit voltage of 0.47 V from the CNT/N719 composite fiber is much lower than the planar DSC.

In order to utilize the CNT/N719 composite fiber electrode in the fiber-shaped DSC, a thin layer of poly(vinylidene fluoride) (PVDF) film with porous structure is coated onto the outer surface of a pristine CNT fiber, which serves as the counter

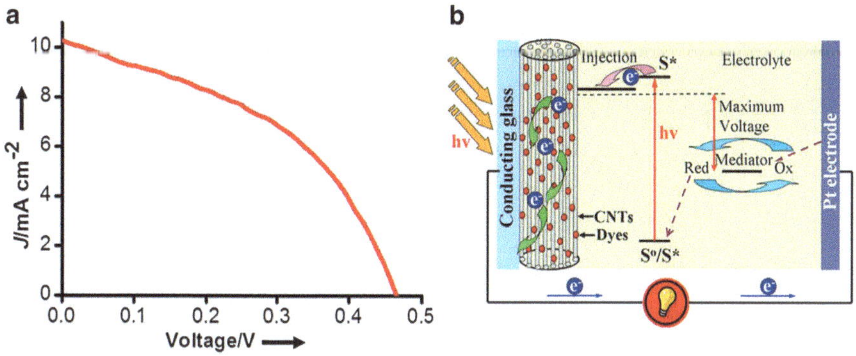

Fig. 3.7 **a** *J–V* curve of a solar cell with an aligned CNT/N719 composite fiber as working electrode. **b** Schematic illumination of the production and transportation of photoelectron in the solar cell with the CNT/N719 composite fiber working electrode (Reproduced from Ref. [39] by permission of John Wiley & Sons Ltd)

electrode [40]. The PVDF film can effectively prevent short circuit, which is caused by direct contact of fiber electrodes. A fiber-shaped DSC is successfully constructed by twining the two fiber electrodes together. Although the efficiency is relatively low, it is a really fiber-shaped DSC fabricated from the CNT fiber working electrode.

3.3.1.2 Carbon Nanotube Fiber Coated with TiO_2 Nanoparticle

Just like the working electrode of planar DSC, TiO_2 was coated onto the FTO glass, and the fiber working electrode was prepared by incorporating the TiO_2 nanoparticles into the CNT fiber [41]. Typically, a CNT/TiO_2 composite fiber was prepared by repeatedly dipping a pristine CNT fiber into a TiO_2 colloid solution, followed by sintering at 500 °C for 60 min. As shown in Fig. 3.8a, b, TiO_2 particles were attached on the outer surface of the CNT fiber, and the diameter of the particle is ~25 nm (Fig. 3.8c, d). The thickness of TiO_2 layer can be controlled by the dipping–coating cycles.

As shown in Fig. 3.9a, b, one CNT fiber coated with a TiO_2 layer was used as working electrode, while the other pristine CNT fiber served as the counter electrode [41]. The two fiber electrodes were tightly twined together to form the desired fiber-shaped DSC (Fig. 3.9c, d). Remarkably, the twisting structure is vital to the successful assembly of the device. Two fiber electrodes can easily contact to cause a short circuit if they are twined too tightly, while the efficiency was severely reduced if two fiber electrodes were wrapped too loosely. A high-power conversion efficiency of 2.94 % was obtained from this fiber-shaped DSC. Figure 3.9e shows that this flexible wire cell can be integrated into a textile by a conventional weaving technology, which is suitable for various portable devices in daily life.

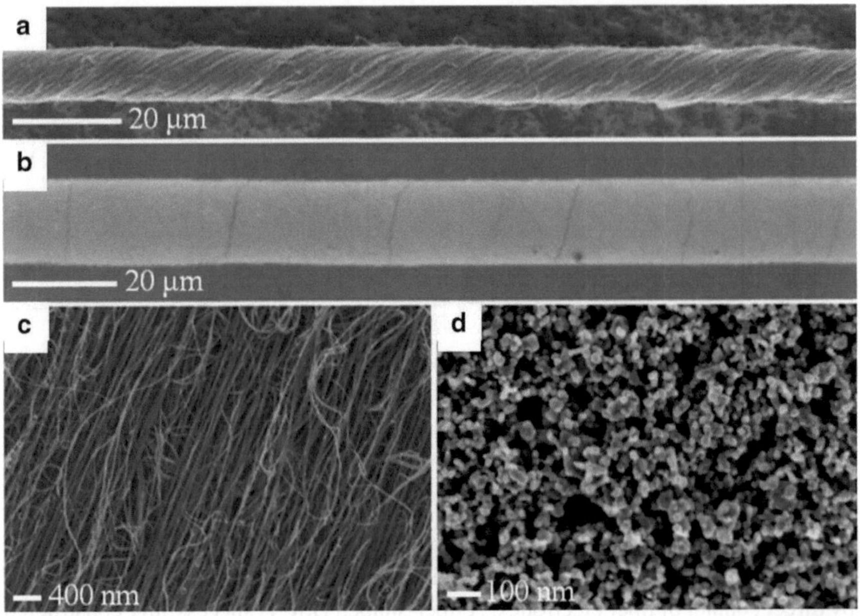

Fig. 3.8 **a** and **b** SEM images of a CNT fiber before and after coating the TiO_2 particles, respectively. **c** and **d** High-resolution images of **a** and **b**, respectively (Reprinted with the permission from Ref. [41]. Copyright 2012 American Chemical Society)

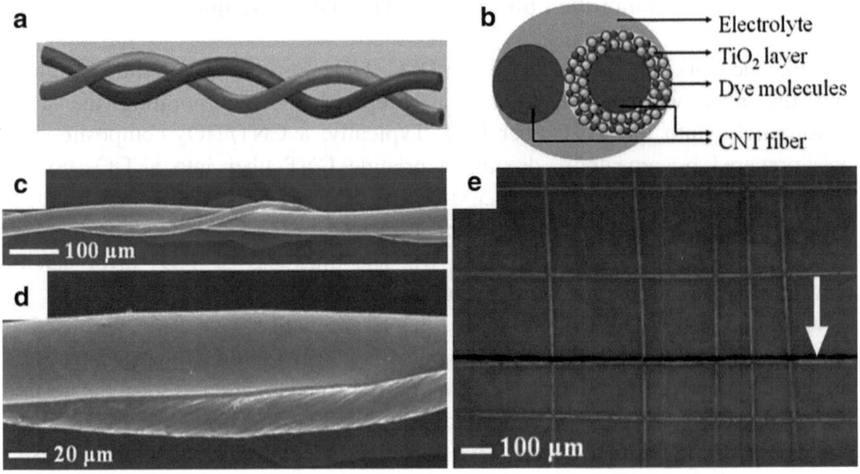

Fig. 3.9 Fiber-shaped DSC based on a CNT fiber coated with TiO_2 particles as working electrode and a pristine CNT fiber as counter electrode. **a** and **b** Schematic illustrations from side and top views, respectively. **c** and **d** SEM of the DSC under different magnifications. **e** A wire cell being woven into a textile composed of aramid fiber. The *arrow* shows the fiber-shaped DSC (Reprinted with the permission from Ref. [41]. Copyright 2012 American Chemical Society)

3.3.1.3 Ti Wire Grown with Aligned TiO_2 Nanotube Array

Although the CNT fiber-based working electrode can successfully serve as fiber working electrode, the cell efficiencies are lower than expected. To find a more suitable fiber working electrode may be a critical step in constructing the high-efficiency fiber-shaped DSC. A layer of three-dimensional TiO_2 nanoparticles film was usually used as an electron transport medium. However, the mobility of electrons in traditional nanoparticles is lower than the one-dimensional nanotube and nanowire by several orders of magnitude in traditional planar DSCs [24, 42]. Hence, the TiO_2 nanotube array can be also used as the candidate for enhancing the electron transport speed, which is beneficial for electron collecting.

Some attempts are very useful for obtaining the TiO_2nanotube-modified fiber working electrode. Typically, growth of aligned TiO_2 nanotubes in the titanium wire is made by electrochemically anodizing a pure titanium wire in the NH_4F/H_2O/ethylene glycol mixed solution [43]. As shown in Fig. 3.10a, the aligned TiO_2 nanotubes were grown on the outer surface of a titanium wire. The diameter of nanotubes ranged from 70 to 100 nm with a wall thickness of 25 nm from top and side views, respectively (Fig. 3.10b–d).

The voltage of anodization plays a critical role on the energy conversion efficiency [43]. As shown in Fig. 3.11a, the anodization at 60 V provides the wire cell with the highest energy conversion efficiency. The concentration of H_2O in the

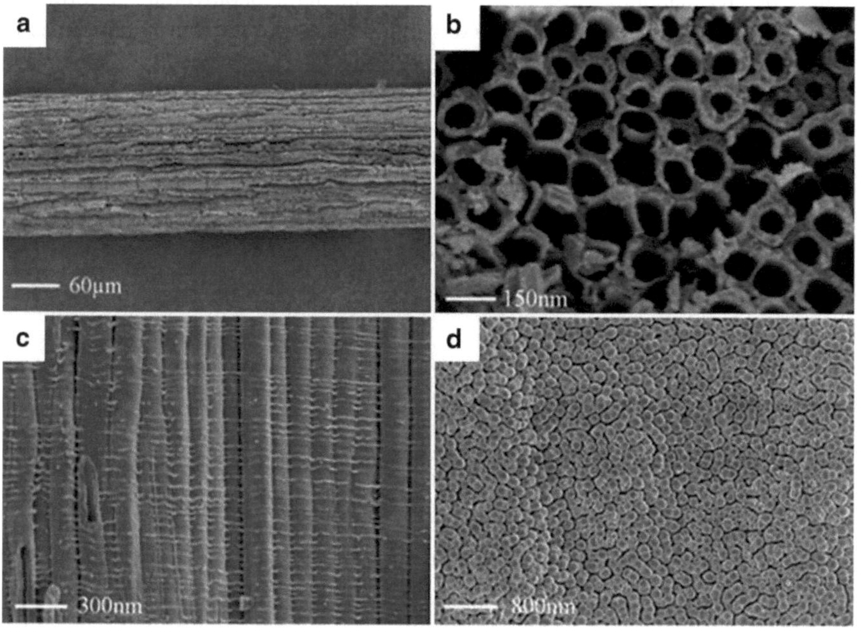

Fig. 3.10 **a** SEM image of a Ti wire after anodized. **b** *Top* view of aligned TiO_2 nanotubes. **c** *Side* view of aligned TiO_2 nanotubes. **d** *Bottom* view of aligned TiO_2 nanotubes after removal from the titanium wire (Reproduced from Ref. [43] by permission of John Wiley & Sons Ltd)

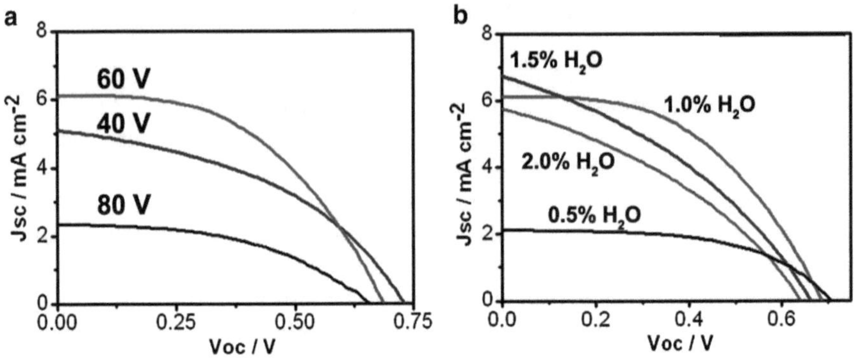

Fig. 3.11 **a** *J–V* curves of fiber-shaped DSCs by using TiO_2nanotube-modified titanium wires being anodized at different voltages for 2 h in the electrolyte containing 1 % of H2O. **b** *J–V* curves of fiber-shaped DSCs by using TiO_2nanotube-modified titanium wires being anodized in the electrolyte with different concentrations of H_2O at 60 V for 2 h (Reproduced from Ref. [43] by permission of John Wiley & Sons Ltd)

electrolyte is also important to the photovoltaic performance. As shown in Fig. 3.11b, when the concentration of H_2O rises from 0.5 to 2 % under the same condition, the maximal efficiency occurred at a H_2O concentration of 1 %. The length of TiO_2 nanotubes also plays a critical role in the photovoltaic performance as it determines the quantity of loaded dye molecules and the transport distance of electrons from the nanotubes to the Ti wire. The highest efficiency can be obtained when the length of TiO_2 nanotube was ~35 μm at the anodizing time of 8 h. Note that these parameters could be changed with different anodization electrolytes and materials. Compared with the CNT fiber-based working electrode, the cell efficiency based on Ti wire modified with TiO_2 nanotubes can achieve 3.94 %. The fiber working electrode based on Ti wire may be a suitable material for the fiber-shaped DSC from the above investigations.

3.3.2 Counter Electrode

Besides the fiber working electrode, the fiber counter electrode is another key factor for high efficiency. Typically, the fiber counter electrode should be flexible with high electrocatalytic activity and electrical conductivity. To enhance the reduction of the triiodide and reduce the recombination of the photoelectrons, the counter electrode should intimately contact with the working electrode and the thread pitch should be small, which is kinetically favorable to the reduction of triiodide. However, the densely wound electrode will shelter dyes from illumination, which is unfavorable for generating photocurrent. As a result, it should strike a balance when twisting the two electrodes. The twisted structure makes the fiber-shaped DSCs flexible and sustainable to deformation during use. A lot of efforts are devoted to designing and discussing twisted fiber-shaped DSCs with different fiber counter electrodes.

3.3.2.1 Carbon Nanotube Fiber

In the conditional planar DSC, platinum is used as efficient counter electrode [44, 45]. However, there remain several disadvantages that limited their future applications, such as their limited source on the earth. Therefore, new electrode materials with high efficiency and low cost are highly desired.

Due to their unique structures and excellent electrical properties, CNTs may be a promising candidate. The random dispersion of CNTs largely decreases their capabilities for charge separation and transport from previous reports. Aligned CNT sheets with good transparency, high flexibility, and excellent electronic properties are very suitable for electrode applications [46–51]. For instance, planar DSCs with aligned CNT sheet and platinum as counter electrode were compared under the same experimental conditions. The resulting DSC with CNT sheet exhibited a V_{OC} of 0.67 V, J_{SC} of 10.80 mA cm^{-2}, and FF of 0.57. The energy conversion efficiency is calculated to be 4.18 %. When the CNT sheets with thickness of 1.20 mm are used as counter electrode, the resulting cell showed an efficiency of 6.60 % compared with 5.27 % of the platinum-based cell [52].

The above features show that CNTs are very efficient counter electrode in DSC; however, use of the CNTs in fiber-shaped DSCs should convert the planar CNT film into macroscopic fiber. Some attempts for using the CNT fiber as counter electrode in fiber-shaped DSC are summarized below [43]. As shown in Fig. 3.12a, a fiber-shaped DSC was fabricated by twisting a CNT fiber around the TiO_2nanotube-modified titanium, and the thread pitch was ~1.8 mm. As shown in Fig. 3.12b, due to the flexibility of the CNT fiber, two electrodes are closely contacted for effective charge separation and transport. Similar to the previous discussion, the pitch distance affects the cell efficiency. As shown in Fig. 3.12c, three fiber-shaped DSCs with different pitches are discussed. Obviously, the J_{SC} at a thread pitch of 0.8 mm is obviously higher than that of 1.4 mm, while the V_{OC} and FF are almost the same. When the two electrodes are parallel in assembling method, the lowest efficiency is obtained due to a poor contact. Note that the optimized parameter may be alternated for CNT fibers with different diameters. The fiber-shaped DSCs with different lengths are also investigated. As shown in Fig. 3.12d, the photovoltaic parameters had not been obviously changed when the cell length varies in 0.2–1.4 cm, and the efficiency is almost the same, which is important for the practical application at large scale.

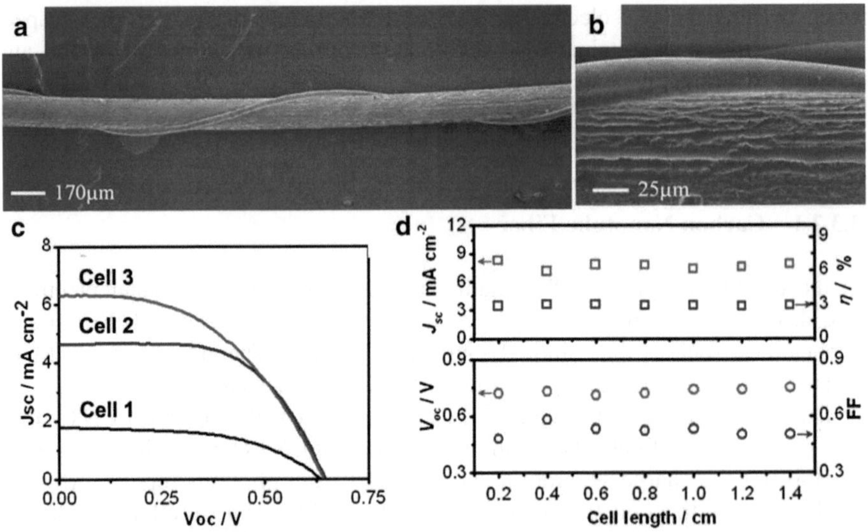

Fig. 3.12 **a** SEM image of the fiber-shaped DSC by twisting a CNT fiber around the TiO_2-nanotube-modified Ti wire. **b** SEM images of a twined part under high magnification. **c** *J–V* curves of the fiber-shaped DSCs with different thread pitches. Two electrodes are parallel for cell 1 and thread pitches are 1.4 and 0.8 mm for cells 2 and 3, respectively. The Ti wire was anodized at 60 V for 2 h in the electrolyte containing 1 % H_2O. **d** Dependence of photovoltaic parameters on the length of fiber-shaped DSCs. The used Ti wires were anodized in the electrolyte containing 1 % of H_2O at 60 V for 8 h (Reproduced from Ref. [43] by permission of John Wiley & Sons Ltd)

3.3.2.2 Carbon-Nanostructure Fiber

Graphene, a single atomic layer of graphite, has attracted increasing interest owing to its remarkable properties [53–56]. However, it is difficult to use the graphene for fiber-shaped DSCs unless one-dimensional graphene fiber like the CNT fiber is obtained. So, it is necessary to prepare graphene materials in the form of a fiber. Typically, the graphene fiber was obtained by the following steps: Firstly, the graphene oxide (GO) was synthesized by a modified Hummer's method. Secondly, GO fibers were fabricated through a wet-spinning method by injecting GO dispersions into a coagulation bath (Fig. 3.13a). Lastly, the graphene fibers were produced by reduction of GO fibers in the aqueous solution of hydroiodic acid at 80 °C [57].

The graphene fiber exhibited a density of 0.61 g cm^{-3}, an electrical conductivity of 10^2–10^3 S cm^{-1}, and a tensile strength of 10^2–10^3 MPa. Figure 3.13b showed the SEM image of a uniform graphene fiber. Its highly stacked structure can be confirmed by the cross section of a broken graphene fiber (Fig. 3.13c). Due to its good flexibility and high mechanical strength, the graphene fiber can even be made into a tight knot (Fig. 3.13d). Those characteristics indicate that the graphene fiber may be useful fiber electrodes for fiber-shaped devices.

The fiber-shaped DSC was fabricated by twisting a graphene fiber around a modified Ti wire [57]. The cell obtains a V_{OC} of 0.72 V, J_{SC} of 12.67 mA cm^{-2}, and

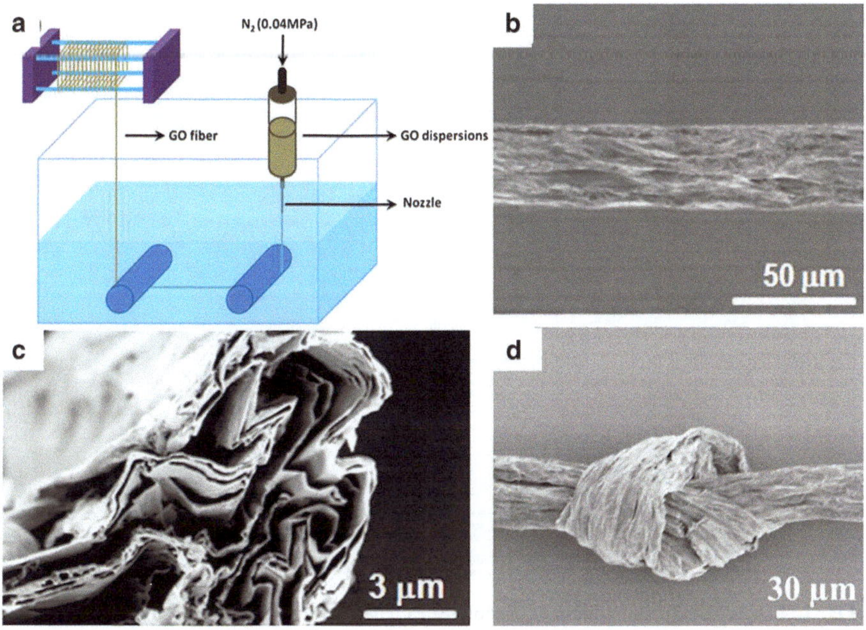

Fig. 3.13 **a** Schematic illustration to the preparation of graphene oxide fibers. **b** SEM image of a graphene fiber. **c** Cross-sectional view of a graphene fiber after breaking. **d** SEM image of a graphene fiber being made into a tight knot (Reproduced from Ref. [57] by permission of John Wiley & Sons Ltd)

FF of 0.42, and the efficiency is 3.85 %. The pristine graphene fibers are not robust as expected for fiber-shaped DSCs. There are some reasons for this result, such as lower electrical conductivity. In order to enhance the efficiency, other components should be incorporated into the graphene fiber.

As an emerging branch of the carbon nano-family, graphene-nanoribbons (GNRs) have aroused attention for various applications such as photovoltaic devices [58–60]. However, it also remains a trouble to assemble GNRs into macroscopical fibers. Incorporating the GNRs into other fibers to form a composite fiber is a useful approach to realize the application of GNRs.

Recently, some attempts are made by unzipping the outer CNTs of CNT fibers into a novel CNT–graphene core–sheath fiber [61]. Figure 3.14a and b shows the schematic illustration and SEM image of the core–sheath nanostructured fiber, respectively, and the corresponding magnified image was shown in Fig. 3.14c. The CNT core was verified after peeling off the carbonaceous sheath in the core–sheath nanostructured fiber (Fig. 3.14d). The fiber-shaped DSCs were fabricated by twisting the core–sheath fiber around a TiO_2nanotube-modified Ti wire. A high efficiency of 5.16 % was achieved based on this core–sheath fiber, which is attributed to the GNRs and carbonaceous residues at the outer sheath with abundant edges directly contacted with the electrolyte.

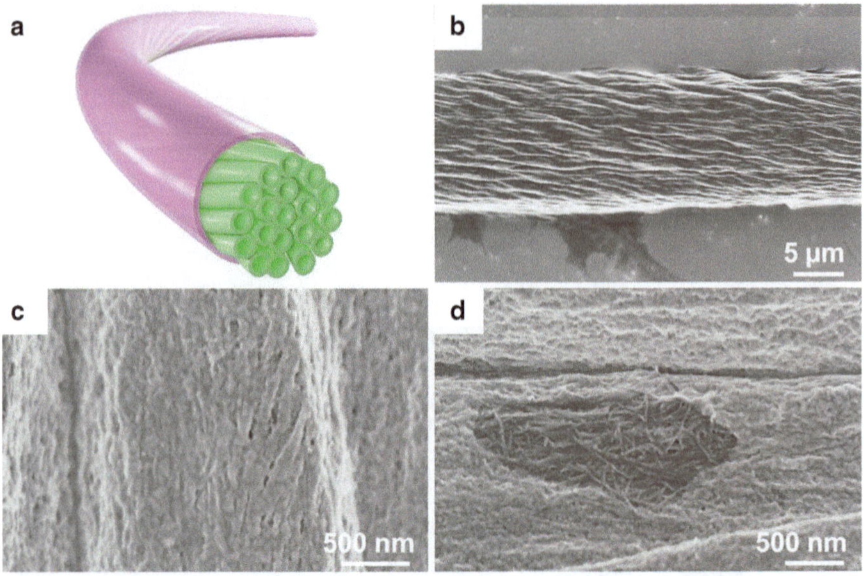

Fig. 3.14 **a** Schematic illustration of the core–sheath nanostructured fiber. **b** and **c** SEM images of an oxidized CNT fiber under different magnifications. **d** SEM image of a core–sheath nanostructured fiber with a sheath part peeled off to unveil the CNTs (Reproduced from Ref. [61] by permission of John Wiley & Sons Ltd

3.3.2.3 Carbon/Platinum Composite Fiber

Although the carbon-based fibers are successfully used for fiber-shaped DSCs, the cell efficiencies are not as high as expected [62]. It is necessary to introduce other components to further enhance the efficiency. Platinum is the first candidate due to their excellent electrochemical activity as previously reported. As mentioned above, a bare graphene fiber is not an efficient counter electrode. To increase the catalytic activity, Pt nanoparticles were electrodeposited by a double potential step method to prepare graphene/Pt composite fibers at room temperature with the first step at E_1 of −0.5 V for 5 s and the second step at E_2 of 0 V for the time range of 10–500 s. The electrolyte contains 5 mM H_2PtCl_6 and 0.5 M H_2SO_4 in aqueous solution [57].

Amazedly, when the graphene fiber was electrodeposited with platinum nanoparticles, the electrocatalytic and electrical conductivity was obviously increased, such as 1,024 S cm^{-1} for 22.9 % Pt. The efficiency of the fiber-shaped DSC based on a Pt/graphene fiber and a titanium wire modified with titania nanotubes was about 8.45 %, which is the recorded efficiency for the fiber-shaped DSCs. In contrast, Pt wire, carbon fiber, and carbon nanotube fiber were compared to electrodeposite with platinum nanoparticles. As shown in Fig. 3.15, the Pt nanoparticles with an average diameter of 8 nm were uniformly dispersed in the graphene fiber, while the nanoparticle easily aggregated to form much larger

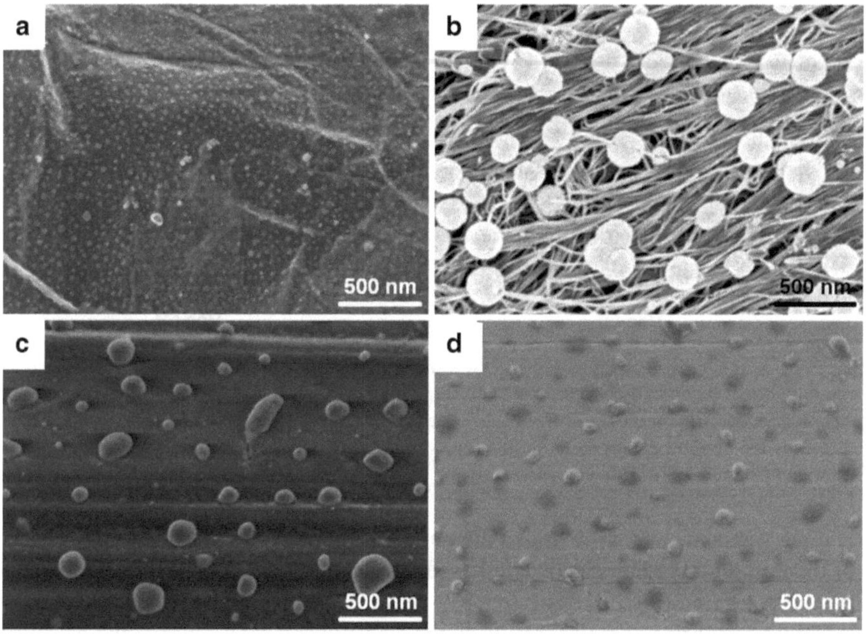

Fig. 3.15 Comparison among four different fibers after electrodeposition of the same content of Pt particles. **a** Graphene fiber after electrodeposition. **b** CNT fiber after electrodeposition. **c** Carbon fiber after electrodeposition. **d** Pt wire after electrodeposition (Reproduced from Ref. [57] by permission of John Wiley & Sons Ltd)

nanoparticles on the surface for the other three fibers. For the graphene fiber, the consecutive carbon nanomaterial provides excellent electronic properties and the Pt nanoparticle gives rise to high electrocatalytic activity.

3.3.3 Electrolyte

The electrolyte is another key component in the DSC. The discussed fiber-shaped DSCs were mainly fabricated on the basis of liquid I^-/I_3^- electrolyte. Although this conventional redox couple works efficiently, disadvantages such as corrosion of the current collector and partial absorption in the visible region will limit their application. Therefore, it is important to study alternative redox couples [63–65]. Due to the negligible absorption in the visible region, thiolate/disulfide (T^-/T_2) organic redox couple was introduced into the fiber-shaped DSC. The UV–vis spectra of the T^-/T_2 and I^-/I_3^- redox couple obviously show that this organic electrolyte has less absorption in the visible region, which is widely known to be important for efficient DSCs.

Fiber-shaped DSCs constructed from this organic redox couple were compared for CNT fibers with different diameters. As shown in Fig. 3.16a, the highest

shaped DSC with electrodes twisted is annoyed by the problem that the electrolyte is readily to evaporate in the absence of glass cladding. The loss of electrolyte makes the cell fade rapidly when exposed in air. The wrapped CNT film, though, is beneficial for retaining the electrolyte inside and retards the evaporation, indicating a better long-suffering capability.

An improvement was made by replacing the CNT film with aligned CNT sheet (Fig. 3.17b). The aligned CNT sheets are also flexible and transparent while with higher electrical conductivities. Likewise, a Ti wire modified with TiO_2 nanotubes was employed as working electrode. The effect of the length of TiO_2 nanotubes and the thickness of CNT sheet were, respectively, discussed. As previously mentioned, longer TiO_2 nanotubes have larger surface areas to absorb more dye molecules and generate higher photoelectron currents. However, when the length further increases, the charge recombination looms large due to the longer diffusion length, which reduces the collection of electrons. As for different thicknesses of CNT sheets, the open-circuit voltage and fill factor were slightly changed, while the short-circuit current density intensively relied on the thickness of the CNT sheet. Since the thicker sheet diminishes the illumination intensity, dye molecules are not fully excited, leading to a lower photoelectron current.

An optimized efficiency of 4.1 % was achieved compared with 3.20 % of the twisted fiber-shaped DSC under the same condition (Fig. 3.17c). The coaxial DSC exhibited a lower J_{SC} than the twisted DSC wire, ascribing to the inferior conductivity of the CNT sheet. It should be noted that the fill factor in coaxial DSC is higher than that of a twisted DSC wire, benefited from the intimate contact between the electrolyte and electrode.

Concerning the vulnerability of liquid electrolyte, further attempt was devoted to adopting eutectic melts to prepare a quasi-solid-state coaxial DSC fiber [70]. Eutectic melts of 1-ethyl-3-methylimidazolium iodide (EMII), ionic liquid of 1-propyl-3-methylimidazolium iodide (PMII), and iodine were mixed and used as the quasi-solid-state electrolyte. Imidazole-like cation is anchored on the surface of working electrode as a Helmholtz-like layer, which inhibits the recombination of I_3^- ions and electrons in the TiO_2 conducting band. The highest energy conversion efficiency of 2.6 % was yielded. The resulting electrolyte is sustainable to temperature, which makes the DSC fiber durable. In addition, these fiber-shaped DSCs could be operated at relatively high temperatures. They were found to be stable, e.g., the eutectic melt-based DSC could be efficiently performed over 300 h in a wide temperature window from 20 to 120 °C.

Another coaxial structure that integrated optical fibers and nanowire arrays as three-dimensional (3D) DSCs was shown in Fig. 3.17d [71]. The use of optical fiber was inspired by the fact that for traditional fiber-shaped DSCs, the entangled nanotubes on the working electrode have a shadow effect to decrease the harvest of light. The primary part of the coaxial fiber consists of a bundle of quartz fibers conducting the incident light. The coaxial fiber can be divided into two segments along axial. One with a cladding low-refractive-index layer is used to conduct light, while the other part is fabricated into the DSC. The as-prepared fiber separates the light incidence and the light harvest, which enable the DSC to be operated in the absence of incident light. The segment for a DSC, without a cladding layer, is superficially coated with an ITO layer instead. The ITO layer simultaneously acts as a conductive electrode and a high-refractive-index material that refracts light out of the fiber. Aligned zinc oxide (ZnO) nanowires are grown from a ZnO layer deposited on the ITO layer as a seed layer, followed by sensitizing with dye. A Pt-deposited glass substrate is used as the counter electrode. Afterwards, the working electrode fiber with sensitized ZnO nanowires is assembled in parallel with the counter electrode. The internal space of the device is injected with a liquid electrolyte. When light enters and travels along the optical fiber after repeated internal reflections, it will be refracted across the interface of ITO/ZnO nanowires and harvested by dye molecules. Internal reflection within the fiber creates multiple opportunities for energy conversion at the interface, rendering an improved conversion efficiency of 3.3 %. Furthermore, the coaxial 3D DSC can be integrated with a nanogenerator to construct a self-powered nanosystem.

3.5 Multifunctional Fiber-Shaped Solar Cell

Fiber-shaped DSCs are flexible and light weight compared with the conditional planar DSCs. In order to cater different field applications, kinds of functionalities are further introduced into them. In particular, when the fiber-shaped DSCs are woven into fabrics, they are needed to be stretchable besides the flexible and lightweight properties. For future use in portable electronic facilities such as automobile and spacecraft, the photovoltaic devices need to be able to be removed as a physical detachment. So, it is vital for fiber-shaped DSCs to have new functionalities besides generating current.

The fiber-shaped DSCs with a functionality in response to magnetic field were realized by incorporating superparamagnetic nanoparticles into the counter electrode. For preparation of the CNT composite fiber, CNT sheets were pulled out of the array and then through the superparamagnetic nanoparticle dispersion [72]. As shown in Fig. 3.18a, the Fe_3O_4/CNT composite fiber is uniform in diameter and its surface is smooth. From the mapping image, the Fe_3O_4 nanoparticles are uniformly dispersed in the CNTs (Fig. 3.18b).

The superconducting quantum interference device was used to investigate the magnetic properties of CNT composite fibers. As shown in Fig. 3.18c, a superparamagnetic behavior was determined by the zero coercivity and remanence on the hysteresis loop. The magnetic property of the composite fibers was shown in Fig. 3.18d, e. When a magnet was close to the composite fiber, it was rapidly attracted to the magnet. As expected, the fiber-shaped DSCs based on this composite fiber twisted around a modified Ti wire could also be stably attached on a magnet. Figure 3.18f, g has compared fiber-shaped DSCs without and with Fe_3O_4 nanoparticles in the counter electrode. Obviously, the bottom DSC with the use of

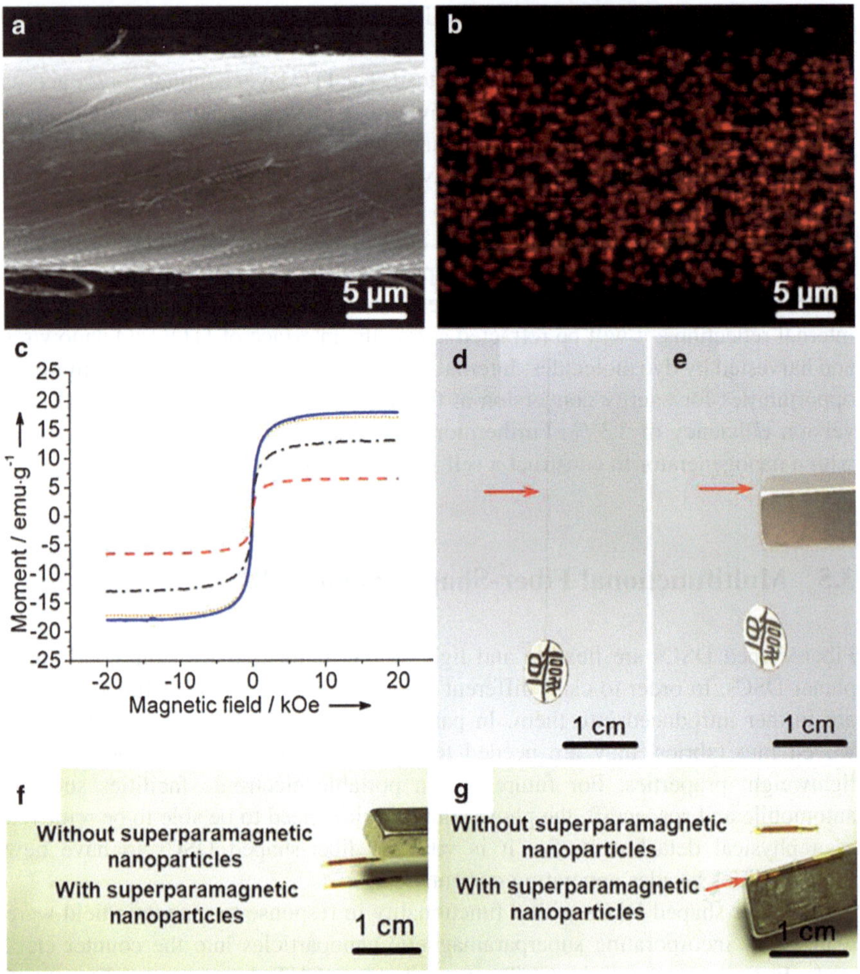

Fig. 3.18 **a** and **b** SEM image and Fe mapping image of a Fe3O4/CNT composite fiber, respectively. **c** Hysteresis loop of the composite fibers with different weight percentages of Fe3O4 nanoparticles. **d** and **e** Optical images of a superparamagnetic nanoparticle/CNT composite fiber before and after being close to a magnet, respectively. **f** and **g** Optical images of two fiber-shaped DSCs before and after a magnet was moved close to them at the same distance. The top device was fabricated by twisting a pristine CNT fiber around a modified titanium wire, while the bottom one was fabricated by twisting a superparamagnetic nanoparticle/CNT composite fiber around the same modified titanium wire (Reproduced from Ref. [72] by permission of John Wiley & Sons Ltd)

magnetic nanoparticles was attracted onto the magnet. After Pt nanoparticles were deposited onto the composite fiber, the efficiency can be enhanced to 8.03 %.

The stretchable property is also very important for electronics to meet different kinds of applications. For the fiber-shaped DSCs to realize stretchable property, they must have a stretchable working and counter electrode. To realize a

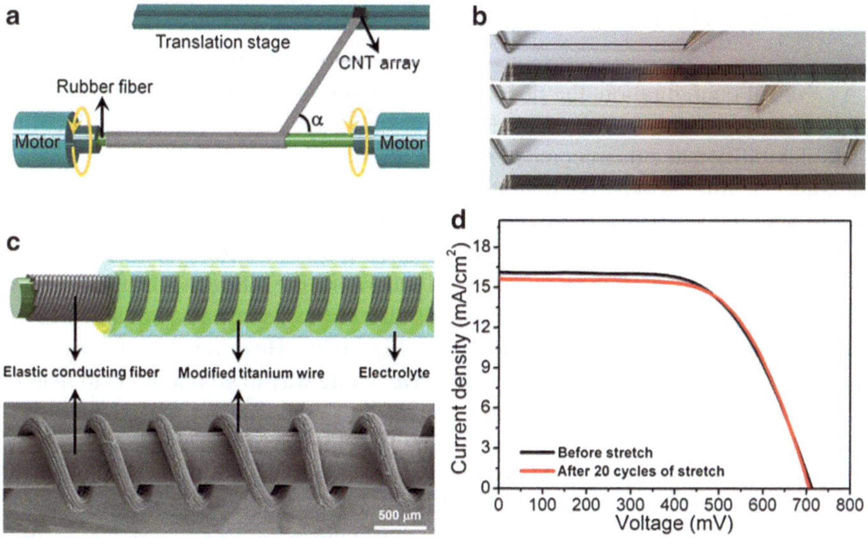

Fig. 3.19 **a** Schematic illustration of the preparation of elastic conducting fibers. **b** An elastic conducting fiber before and after stretching by 50 % and 100 %, respectively. **c** Schematic illustration and SEM image of a stretchable DSC with pitch distance of 560 μm. **d** *J–V* curves of a stretchable DSC before and after stretching for 20 cycles at a strain of 30 % (Reproduced from Ref. [74] by permission of John Wiley & Sons Ltd)

stretching property, a mini-3D DSC based on a spiral-shaped titanium wire was designed [73]. The device could easily hold liquid electrolyte due to the capillary force, and an efficiency of 4.1 % was obtained. Although the device cannot be stretchable, the spiral-shaped structure gives researchers some enlightenment. The spiral-shaped titanium wire may work like a spring, so it can be stretched after modifications.

The fiber counter electrode should be elastic for the stretchable fiber-shaped DSC. Besides the stretchable property, the elastic fiber electrode should also have high electrocatalytic activity. Such an elastic conducting fiber was prepared by winding aligned CNT sheets on a rubber fiber (Fig. 3.19a). These stretchable fibers were also flexible and could be knotted into various forms. No obvious damages were observed in structure after stretching by 100 % for many cycles (Fig. 3.19b). Moreover, the electrical resistance of the fiber remained almost unchanged after stretching.

To fabricate the fiber-shaped stretchable DSC, a spring-like Ti wire was grown with a layer of aligned TiO_2 nanotubes, and the elastic conducting fiber is inserted into the spring-like Ti working electrode (Fig. 3.19c) [74]. At last, both electrodes were transferred to a stretchable, transparent polyethylene tube. The maximal efficiency of 7.13 % was obtained with a pitch distance of the twisted working electrode to be 560 μm. As shown in Fig. 3.19d, the *J–V* curves of a stretchable DSC were almost identical before and after stretching by 30 % for 20 cycles.

3.6 Perspective

As a novel configuration, the fiber-shaped DSC represents a cutting-edge technology in energy conversion devices. Merits of small size, light weight, high ductility, and flexibility make them competent to be integrated into miniature devices. Their unique one-dimensional configurations provide opportunities to weave an energy cloth with the assistance of textile technologies.

Reviewing the development of fiber-shaped DSCs in a few years, we shall be surprised by its promotion in performance and promising future in application. From less than 1 % to over 8 %, the revolution of fiber electrodes always provides a strong impetus to the elevation of performances (Fig. 3.20). However, haunted by some tricky issues, the fiber-shaped DSCs still have a long way to go out of lab to market. Compared with their counterparts, the planar format, the energy conversion efficiencies are not competitive yet. The reported highest efficiency is 8.45 %, less than 10 %, comparable but not enough. Moreover, the lengths of the conceptually developed fiber-shaped DSCs are within the range from several to tens of centimeters, which is not eligible for textile. And if the cells are fabricated into a large size, their efficiencies slump as the increasing resistances of fiber electrodes. The fiber-shaped DSCs with high conversion efficiency and reasonable lengths are badly needed. One of the promising approaches towards this problem is developing high-performance fiber electrodes integrated with high electrical conductivities and catalytic activities.

Electrolyte is another problem that impedes their ubiquity. The volatility and thermal instability of the liquid electrolytes make the cell vulnerable during long-time use, which looms large for the fiber-shaped DSCs, because of encapsulation difficulties. Further endeavor should focus on the DSCs adopting solid-state electrolytes. In this case, polymer solar cells are more competent for fiber-shaped photovoltaic devices since the involved materials are always in solid state, but their superior conversion efficiencies cannot rival those of DSCs.

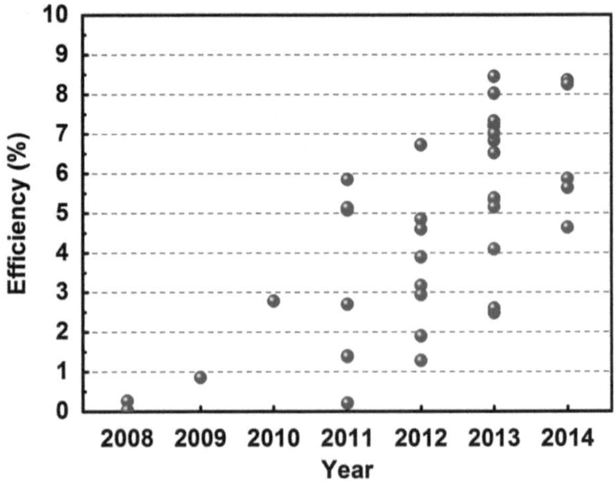

Fig. 3.20 The reported conversion efficiency for fiber-shaped DSCs

The good news is that a scaling-up attempt, that is, weaving the fiber-shaped solar cells into an energy textile, was reported recently (Chap. 9). A metal–textile electrode prepared from micrometer-sized metal wires was employed as a working electrode, and the CNT fibers were woven into a textile as counter electrode. The produced DSC cloth delivered a decent energy conversion efficiency and exhibited high endurance to deformation, which promises an up-and-coming application pertaining to wearable devices.

References

1. McEvoy A, Grätzel M (1994) Sensitisation in photochemistry and photovoltaics. Sol Energy Mater Sol Cells 32(3):221–227
2. Desilvestro J, Graetzel M, Kavan L, Moser J, Augustynski J (1985) Highly efficient sensitization of titanium dioxide. J Am Chem Soc 107(10):2988–2990
3. O'regan B, Grfitzeli M (1991) A low-cost, high-efficiency solar cell based on dye-sensitized. Nature 353:737–740
4. Hagfeldt A, Graetzel M (1995) Light-induced redox reactions in nanocrystalline systems. Chem Rev 95(1):49–68
5. Hagfeldt A, Boschloo G, Sun L, Kloo L, Pettersson H (2010) Dye-sensitized solar cells. Chem Rev 110(11):6595–6663
6. Cahen D, Hodes G, Grätzel M, Guillemoles JF, Riess I (2000) Nature of photovoltaic action in dye-sensitized solar cells. J Phys Chem B 104(9):2053–2059
7. Peter LM (2007) Dye-sensitized nanocrystalline solar cells. Phys Chem Chem Phys 9(21):2630–2642
8. Memming R (1984) Electron transfer process with excited molecules at semiconductor electrodes. Prog Surf Sci 17(1):7–73
9. Grätzel M (2005) Solar energy conversion by dye-sensitized photovoltaic cells. Inorg Chem 44(20):6841–6851
10. Rosenbluth ML, Lewis NS (1989) "Ideal" behavior of the open circuit voltage of semiconductor/liquid junctions. J Phys Chem 93(9):3735–3740
11. Sommeling P, O'regan B, Haswell R, Smit H, Bakker N, Smits J, Kroon J, Van Roosmalen J (2006) Influence of a TiCl4 post-treatment on nanocrystalline TiO_2 films in dye-sensitized solar cells. J Phys Chem B 110(39):19191–19197
12. Roy P, Berger S, Schmuki P (2011) TiO_2 nanotubes: synthesis and applications. Angew Chem Int Ed 50(13):2904–2939
13. Roy P, Kim D, Lee K, Spiecker E, Schmuki P (2010) TiO_2 nanotubes and their application in dye-sensitized solar cells. Nanoscale 2(1):45–59
14. Zhu K, Neale NR, Miedaner A, Frank AJ (2007) Enhanced charge-collection efficiencies and light scattering in dye-sensitized solar cells using oriented TiO_2 nanotubes arrays. Nano Lett 7(1):69–74
15. Jennings JR, Ghicov A, Peter LM, Schmuki P, Walker AB (2008) Dye-sensitized solar cells based on oriented TiO_2 nanotube arrays: transport, trapping, and transfer of electrons. J Am Chem Soc 130(40):13364–13372
16. Quintana M, Edvinsson T, Hagfeldt A, Boschloo G (2007) Comparison of dye-sensitized ZnO and TiO_2 solar cells: studies of charge transport and carrier lifetime. J Phys Chem C 111(2):1035–1041
17. Fujishima A, Honda K (1972) Photolysis-decomposition of water at the surface of an irradiated semiconductor. Nature 238(5385):37–38
18. Grätzel M (2001) Photoelectrochemical cells. Nature 414(6861):338–344

19. Baps B, Eber-Koyuncu M, Koyuncu M (2001) Ceramic based solar cells in fiber form. Key Eng Mater 206:937–940
20. Liu J, Namboothiry MA, Carroll DL (2007) Fiber-based architectures for organic photovoltaics. Appl Phys Lett 90(6):063501
21. O'Connor B, Pipe KP, Shtein M (2008) Fiber based organic photovoltaic devices. Appl Phys Lett 92(19):193306
22. Fan X, Chu Z, Wang F, Zhang C, Chen L, Tang Y, Zou D (2008) Wire-shaped flexible dye-sensitized solar cells. Adv Mater 20(3):592–595
23. Chen T, Qiu L, Yang Z, Peng H (2013) Novel solar cells in a wire format. Chem Soc Rev 42(12):5031–5041
24. Wang H, Liu Y, Li M, Huang H, Zhong M, Shen H (2009) Hydrothermal growth of large-scale macroporous TiO_2 nanowires and its application in 3D dye-sensitized solar cells. Appl Phys A 97(1):25–29
25. Liu Z, Misra M (2010) Dye-sensitized photovoltaic wires using highly ordered TiO_2 nanotube arrays. ACS Nano 4(4):2196–2200
26. Lv Z, Yu J, Wu H, Shang J, Wang D, Hou S, Fu Y, Wu K, Zou D (2012) Highly efficient and completely flexible fiber-shaped dye-sensitized solar cell based on TiO_2 nanotube array. Nanoscale 4(4):1248–1253
27. Lv Z, Fu Y, Hou S, Wang D, Wu H, Zhang C, Chu Z, Zou D (2011) Large size, high efficiency fiber-shaped dye-sensitized solar cells. Phys Chem Chem Phys 13(21):10076–10083
28. Fu Y, Lv Z, Hou S, Wu H, Wang D, Zhang C, Chu Z, Cai X, Fan X, Wang ZL (2011) Conjunction of fiber solar cells with groovy micro-reflectors as highly efficient energy harvesters. Energy Environ Sci 4(9):3379–3383
29. Chen L, Zhou Y, Dai H, Li Z, Yu T, Liu J, Zou Z (2013) Fiber dye-sensitized solar cells consisting of TiO_2 nanowires arrays on Ti thread as photoanodes through a low-cost, scalable route. J Mater Chem A 1(38):11790–11794
30. Tao H, G-j F, W-j K, Zeng W, Wang J (2014) In-situ synthesis of TiO_2 network nanoporous structure on Ti wire substrate and its application in fiber dye sensitized solar cells. J Power Sources 245:59–65
31. Ramier J, Da Costa N, Plummer C, Leterrier Y, Månson J-A, Eckert R, Gaudiana R (2008) Cohesion and adhesion of nanoporous TiO_2 coatings on titanium wires for photovoltaic applications. Thin Solid Films 516(8):1913–1919
32. Ramier J, Plummer C, Leterrier Y, Månson J-A, Eckert B, Gaudiana R (2008) Mechanical integrity of dye-sensitized photovoltaic fibers. Renew Energy 33(2):314–319
33. Hou S, Cai X, Fu Y, Lv Z, Wang D, Wu H, Zhang C, Chu Z, Zou D (2011) Transparent conductive oxide-less, flexible, and highly efficient dye-sensitized solar cells with commercialized carbon fiber as the counter electrode. J Mater Chem 21(36):13776–13779
34. Zhang S, Ji C, Bian Z, Yu P, Zhang L, Liu D, Shi E, Shang Y, Peng H, Cheng Q (2012) Porous, platinum nanoparticle-adsorbed carbon nanotube yarns for efficient fiber solar cells. ACS Nano 6(8):7191–7198
35. Cai X, Hou S, Wu H, Lv Z, Fu Y, Wang D, Zhang C, Kafafy H, Chu Z, Zou D (2012) All-carbon electrode-based fiber-shaped dye-sensitized solar cells. Phys Chem Chem Phys 14(1):125–130
36. Cai X, Wu H, Hou S, Peng M, Yu X, Zou D (2013) Dye-sensitized solar cells with vertically aligned TiO_2 nanowire arrays grown on carbon fibers. ChemSusChem 7(2):474–482
37. Yen C-Y, Lin Y-F, Liao S-H, Weng C-C, Huang C-C, Hsiao Y-H, Ma C-CM, Chang M-C, Shao H, Tsai M-C (2008) Preparation and properties of a carbon nanotube-based nanocomposite photoanode for dye-sensitized solar cells. Nanotechnology 19(37):375305
38. Zhang X, Li Q, Tu Y, Li Y, Coulter JY, Zheng L, Zhao Y, Jia Q, Peterson DE, Zhu Y (2007) Strong carbon-nanotube fibers spun from long carbon-nanotube arrays. Small 3(2):244–248
39. Chen T, Wang S, Yang Z, Feng Q, Sun X, Li L, Wang ZS, Peng H (2011) Flexible, light-weight, ultrastrong, and semiconductive carbon nanotube fibers for a highly efficient solar cell. Angew Chem Int Ed 50(8):1815–1819

40. Cai F, Chen T, Peng H (2012) All carbon nanotube fiber electrode-based dye-sensitized photovoltaic wire. J Mater Chem 22(30):14856–14860
41. Chen T, Qiu L, Cai Z, Gong F, Yang Z, Wang Z, Peng H (2012) Intertwined aligned carbon nanotube fiber based dye-sensitized solar cells. Nano Lett 12(5):2568–2572
42. Wang Y, Liu Y, Yang H, Wang H, Shen H, Li M, Yan J (2010) An investigation of DNA-like structured dye-sensitized solar cells. Curr Appl Phys 10(1):119–123
43. Chen T, Qiu L, Kia HG, Yang Z, Peng H (2012) Designing aligned inorganic nanotubes at the electrode interface: towards highly efficient photovoltaic wires. Adv Mater 24(34):4623–4628
44. Yoon CH, Vittal R, Lee J, Chae W-S, Kim K-J (2008) Enhanced performance of a dye-sensitized solar cell with an electrodeposited-platinum counter electrode. Electrochim Acta 53(6):2890–2896
45. Hsieh T-L, Chen H-W, Kung C-W, Wang C-C, Vittal R, Ho K-C (2012) A highly efficient dye-sensitized solar cell with a platinum nanoflowers counter electrode. J Mater Chem 22(12):5550–5559
46. Huang S, Li L, Yang Z, Zhang L, Saiyin H, Chen T, Peng H (2011) A new and general fabrication of an aligned carbon nanotube/polymer film for electrode applications. Adv Mater 23(40):4707–4710
47. Li L, Yang Z, Gao H, Zhang H, Ren J, Sun X, Chen T, Kia HG, Peng H (2011) Vertically aligned and penetrated carbon nanotube/polymer composite film and promising electronic applications. Adv Mater 23(32):3730–3735
48. Huang S, Yang Z, Zhang L, He R, Chen T, Cai Z, Luo Y, Lin H, Cao H, Zhu X (2012) A novel fabrication of a well distributed and aligned carbon nanotube film electrode for dye-sensitized solar cells. J Mater Chem 22(33):16833–16838
49. Sun X, Chen T, Yang Z, Peng H (2012) The alignment of carbon nanotubes: an effective route to extend their excellent properties to macroscopic scale. Acc Chem Res 46(2):539–549
50. Yang Z, Li L, Lin H, Luo Y, He R, Qiu L, Ren J, Peng H (2012) Penetrated and aligned carbon nanotubes for counter electrodes of highly efficient dye-sensitized solar cells. Chem Phys Lett 549:82–85
51. Guan G, Yang Z, Qiu L, Sun X, Zhang Z, Ren J, Peng H (2013) Oriented PEDOT: PSS on aligned carbon nanotubes for efficient dye-sensitized solar cells. J Mater Chem A 1(42):13268–13273
52. Yang Z, Chen T, He R, Guan G, Li H, Qiu L, Peng H (2011) Aligned carbon nanotube sheets for the electrodes of organic solar cells. Adv Mater 23(45):5436–5439
53. Zhang Y, Tan Y-W, Stormer HL, Kim P (2005) Experimental observation of the quantum Hall effect and Berry's phase in graphene. Nature 438(7065):201–204
54. Schedin F, Geim A, Morozov S, Hill E, Blake P, Katsnelson M, Novoselov K (2007) Detection of individual gas molecules adsorbed on graphene. Nat Mater 6(9):652–655
55. Gupta V, Chaudhary N, Srivastava R, Sharma GD, Bhardwaj R, Chand S (2011) Luminescent graphene quantum dots for organic photovoltaic devices. J Am Chem Soc 133(26):9960–9963
56. Kavan L, Yum J-H, Grätzel M (2011) Graphene nanoplatelets outperforming platinum as the electrocatalyst in co-bipyridine-mediated dye-sensitized solar cells. Nano Lett 11(12):5501–5506
57. Yang Z, Sun H, Chen T, Qiu L, Luo Y, Peng H (2013) Photovoltaic wire derived from a graphene composite fiber achieving an 8.45 % energy conversion efficiency. Angew Chem Int Ed 52(29):7545–7548
58. Kosynkin DV, Higginbotham AL, Sinitskii A, Lomeda JR, Dimiev A, Price BK, Tour JM (2009) Longitudinal unzipping of carbon nanotubes to form graphene nanoribbons. Nature 458 (7240):872–876
59. Cai J, Ruffieux P, Jaafar R, Bieri M, Braun T, Blankenburg S, Muoth M, Seitsonen AP, Saleh M, Feng X (2010) Atomically precise bottom-up fabrication of graphene nanoribbons. Nature 466(7305):470–473
60. Yang Z, Liu M, Zhang C, Tjiu WW, Liu T, Peng H (2013) Carbon nanotubes bridged with graphene nanoribbons and their use in high-efficiency dye-sensitized solar cells. Angew Chem Int Ed 52(14):3996–3999

61. Fang X, Yang Z, Qiu L, Sun H, Pan S, Deng J, Luo Y, Peng H (2014) Core-sheath carbon nanostructured fibers for efficient wire-shaped dye-sensitized solar cells. Adv Mater 26(11):1694–1698
62. Pan S, Yang Z, Chen P, Fang X, Guan G, Zhang Z, Deng J, Peng H (2013) Carbon nanostructured fibers as counter electrodes in wire-shaped dye-sensitized solar cells. J Phys Chem C 118(30):16419–16425
63. Wang M, Chamberland N, Breau L, Moser J-E, Humphry-Baker R, Marsan B, Zakeeruddin SM, Grätzel M (2010) An organic redox electrolyte to rival triiodide/iodide in dye-sensitized solar cells. Nat Chem 2(5):385–389
64. Hamann TW, Ondersma JW (2011) Dye-sensitized solar cell redox shuttles. Energy Environ Sci 4(2):370–381
65. Yella A, Lee H-W, Tsao HN, Yi C, Chandiran AK, Nazeeruddin MK, Diau EW-G, Yeh C-Y, Zakeeruddin SM, Grätzel M (2011) Porphyrin-sensitized solar cells with cobalt (II/III)–based redox electrolyte exceed 12 percent efficiency. Science 334(6056):629–634
66. Pan S, Yang Z, Li H, Qiu L, Sun H, Peng H (2013) Efficient dye-sensitized photovoltaic wires based on an organic redox electrolyte. J Am Chem Soc 135(29):10622–10625
67. Wang D, Hou S, Wu H, Zhang C, Chu Z, Zou D (2011) Fiber-shaped all-solid state dye sensitized solar cell with remarkably enhanced performance via substrate surface engineering and TiO_2 film modification. J Mater Chem 21(17):6383–6388
68. Li H, Yang Z, Qiu L, Fang X, Sun H, Chen P, Pan S, Peng H (2014) Stable wire-shaped dye-sensitized solar cells based on eutectic melts. J Mater Chem A 2:3841–3846
69. Zhang S, Ji C, Bian Z, Liu R, Xia X, Yun D, Zhang L, Huang C, Cao A (2011) Single-wire dye-sensitized solar cells wrapped by carbon nanotube film electrodes. Nano Lett 11 (8):3383–3387
70. Sun H, Li H, You X, Yang Z, Deng J, Qiu L, Peng H (2014) Quasi-solid-state, coaxial, fiber-shaped dye-sensitized solar cells. J Mater Chem A 2(2):345–349
71. Weintraub B, Wei Y, Wang ZL (2009) Optical fiber/nanowire hybrid structures for efficient three-dimensional dye-sensitized solar cells. Angew Chem Int Ed 48(47):8981–8985
72. Sun H, Yang Z, Chen X, Qiu L, You X, Chen P, Peng H (2013) Photovoltaic wire with high efficiency attached onto and detached from a substrate using a magnetic field. Angew Chem Int Ed 52(32):8276–8280
73. Liu Y, Li M, Wang H, Zheng J, Xu H, Ye Q, Shen H (2010) Synthesis of TiO_2 nanotube arrays and its application in mini-3D dye-sensitized solar cells. J Phys D Appl Phys 43(20):205103
74. Yang Z, Deng J, Sun X, Li H, Peng H (2014) Stretchable, wearable dye-sensitized solar cells. Adv Mater 26(17):2643–2647

Chapter 4
Fiber-Shaped Polymer Solar Cell

Abstract Currently, fiber-shaped solar cells are materialized in two types of solar cells, the polymer solar cell (PSC) and the dye-sensitized solar cell (DSC), whose structures are technically feasible for transforming into one-dimensional configuration. In this chapter, we discuss the polymer solar cell first including its developing history and working mechanisms. It is widely acknowledged that the electrode plays a pivotal role in the performance of solar cells. On this account, we proceed from remolding different materials into fiber shape as electrodes and shed light on their impact on flexibility, stability, and power conversion efficiency (PCE) of the polymer solar cell. Moreover, we present our attempts towards application pertaining to wearable devices. Specifically, we fabricated the stretchable polymer solar cells and integrated the solar cell with supercapacitors, which is discussed in detail in this chapter.

4.1 Overview of Polymer Solar Cell

Combining polymer with solar cell sounds contrived, but it indeed works. Polymer solar cell (PSC) is an important branch of organic photovoltaic devices that gains increasing importance in harvesting solar energy. But not all the polymers conform to the requirements for solar cells. Conjugated polymers, where electrons are delocalized along the backbone, are essential in polymer solar cells. The popularity of PSC should be attributed to several advantages: thin-film architecture benefited from high absorption coefficients, abundant raw materials that are little consumed in devices, cost-effective and energy-economical fabrication that is readily scaled up, and devisable architecture that offers opportunities for diversifying devices. Moreover, PSCs are in all-solid state which is favorable for applications in flexible, stretchable devices. Comparing with its counterpart, small-molecule organic photovoltaic, PSCs are usually fabricated through a solution process that is compatible with roll-to-roll printing techniques for large-scale production.

The PSC is originated from the organic photovoltaic based on donor–acceptor bilayer planar heterojunction, which emerged in 1979 with a PCE of ~1 % [1]. But until 1993, first planar heterojunction PSC was demonstrated in the wake of the

H. Peng, *Fiber-Shaped Energy Harvesting and Storage Devices*, Nanostructure Science and Technology, DOI 10.1007/978-3-662-45744-3_4

study in photoinduced electron transfer from conducting polymer to fullerene derivatives [2]. PSC have been fiercely progressed in the following years; records of PCE are frequently refreshed with the revolution in materials, device architectures, and fabrication techniques. Recent efforts have pushed PCE to 10.6 % [3]. But it was turned over before long. The efficiency is updated to 11.5 % in a triple-junction PSC [4]. In the first part of this chapter, we will discuss the working mechanism of the PSC.

4.1.1 Working Mechanism

Conducting polymers, with π–π conjugated delocalization system, are generally regarded as intrinsic wide bandgap semiconductors (>1.4 eV) with negligible intrinsic charge carriers in the dark. When they are illuminated, free-charge carriers are generated by a photoinduced electron transfer from donor (p-type semiconductor) to acceptor (n-type semiconductor). A donor-to-acceptor heterojunction is the core part in PSC that dominates the photoexcitation and charge separation which is similar with the p–n junction in traditional silicon-based solar cells. At present, most of high-performance PSCs are based on polymer–fullerene system where conducting polymers act as electron donors and fullerene (C_{60}) and its derivatives act as electron acceptors [5, 6].

The working mechanism of a PSC can be described as follows: when a PSC is illuminated, the donor material absorbing light is excited and generate excitons, associated electron–hole pairs. The excitons diffuse to the donor–acceptor interface and dissociate into free-charge carriers, which then transfer to respective electrodes (Fig. 4.1a) [6, 7].

A semiconducting polymer can be excited provided that incident photons have energy higher than the bandgap. On account of the large bandgap of semiconducting polymer (~2 eV), which corresponds to a small range of spectrum available, narrowing the bandgap is a powerful strategy to enhance the utilization of solar energy (Fig. 4.1b). The excited polymer does not give out free-charge carriers immediately. Instead, electrostatically associated electron–hole pairs, called excitons, are generated. The exciton disassociation, in other words, charge separation, usually takes place at the donor–acceptor interface induced by the potential gradients across the interface. Meanwhile, excitons are prone to decay through radiative or nonradiative ways before they diffuse to the interface. Therefore, the donor and acceptor domains should be on par with exciton diffusion lengths (10–20 nm) (Fig. 4.1c). After that, the free-charge carriers, electrons, and holes are migrating to respective electrodes. The difference of the ionization potential of the excited donor and the electron affinity of the acceptor, except supplying exciton disassociation, provides the initial driving force for charge transfer [8]. As a result, electrons are excited from the highest occupied molecular orbital (HOMO) into the lowest unoccupied molecular orbital (LUMO) of the donor and then transfer to the LUMO of the acceptor while holes transport through the donor material and get collected by

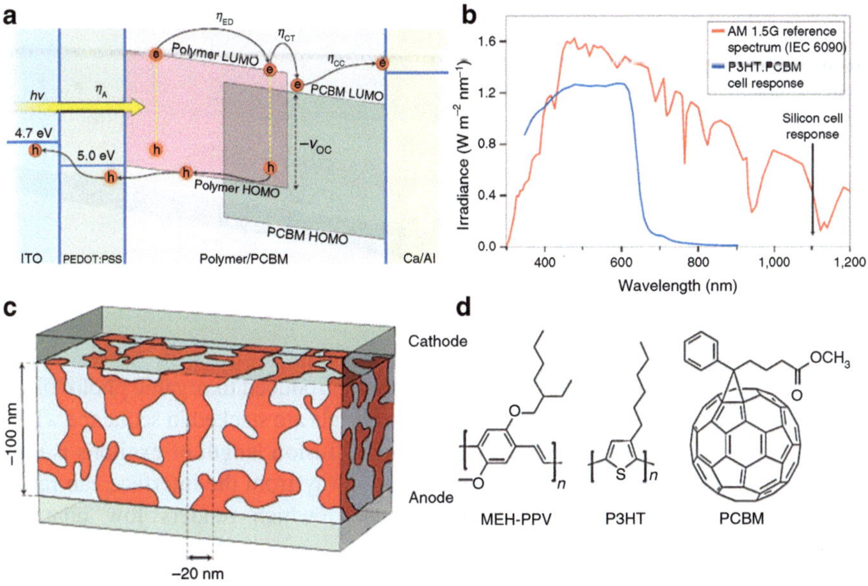

Fig. 4.1 **a** Schematic illustration to the working mechanism of a PSC. **b** Solar spectrum and photoresponse curve of a P3HT:PCBM solar cell. **c** Conceptual morphology model for phase separation of the donor and acceptor. **d** Chemical structures of representative donor and acceptor molecules used in PSCs (Reprinted by permission from Nature Publishing Group: Ref. [6], copyright 2012)

the electrodes. The charge transfer is dominated in competition with photoluminescence and back-transfer recombination since the time charge transfer takes is several orders of magnitude smaller than the latter two processes. In an optimized condition, the photon-to-electron conversion efficiency can approximate to 100 %. But during their journey towards electrodes, recombination looms and affects the charge-collection efficiency [5].

4.1.2 Structure

A typical PSC exhibits a multilayered structure (Fig. 4.2). A transparent indium tin oxide (ITO) glass is used as the hole extracting electrode that is coated with a hole transport layer, like poly(ethylene-dioxythiophene) doped with polystyrene sulfonic acid (PEDOT:PSS). A lower work function metal such as aluminum is used as cathode deposited with a sub-monolayer of lithium fluoride (LiF). Functions of these materials will be elucidated in the next section. Here, we focus our discussion on the intercalated active materials. There are two configurations of donor–acceptor heterojunction in active materials: the planar heterojunction and the bulk heterojunction.

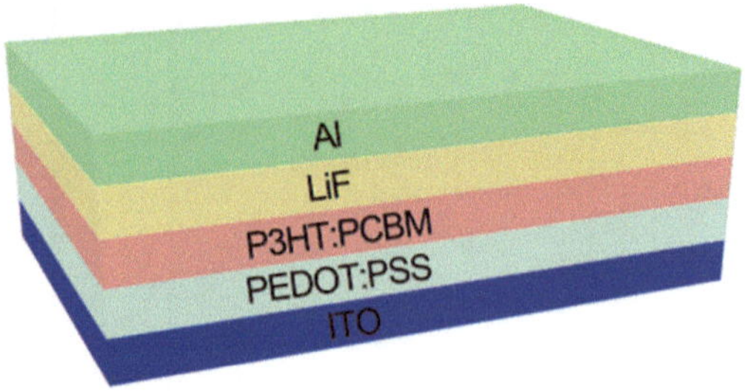

Fig. 4.2 Typical illustration to the multilayered structure of polymer solar cell

4.1.2.1 Planar Heterojunction

The conception of planar heterojunction was first introduced to organic photovoltaic device in 1979 by Tang. Donor and accepter materials are stacked in sequence which forms a planar heterojunction interface. As the diffusion length of excitons is only 10–20 nm, excitons generated more than 20 nm away from the interface can rarely reach the heterojunction and get disassociated, which renders low quantum efficiency.

4.1.2.2 Bulk Heterojunction

The problems that annoy planar heterojunction PSCs are figured out by introducing bulk heterojunction where donor and acceptor materials are blended in a bulk state. Compared with the planar heterojunction where donor and acceptor materials are separated, two phases are interpenetrated in bulk heterojunction. It results in a phase separation in 10–20 nm scale that significantly extends heterojunction interfaces throughout the entire active materials. On this account, bulk heterojunction devices are sensitive to the nanoscale morphology of the photoactive blend. Moreover, phase separation in blend will coarsen during long-term explosion under sunlight leading to morphology degradation, which is also accompanied by photodegradation of conjugated polymer. Hence, the stability of device should be taken into consideration.

4.1.3 Materials

The progress in PSC is always accompanied by the upgrade in materials. In this section, we will cast light on the materials in PSCs, including polymer donor materials, acceptor materials, and electrode materials. In contrast with the diverse donor materials, acceptor materials are limited within fullerene and its derivatives. [6, 6]-Phenyl-C_{61}-butyric acid methyl ester, denoted as PCBM, a more soluble fullerene derivative, is the most important and widely used electron acceptor in PSCs due to its electronegativity and high electron mobility (~10^{-3} cm^2 V^{-1} s^{-1}) (Fig. 4.1d) [5]. Polymer–fullerene systems now take the leading position in high-efficiency PSCs.

4.1.3.1 Polymer Donor Materials

The word "polymer" in the name of polymer solar cell refers to the polymer donor material, which plays a pivotal role in a PSC. An ideal polymer donor material should satisfy several requirements:

1. A narrow bandgap to ensure a broad-spectrum sensitivity
2. Suitable HOMO/LUMO energy levels for efficient charge transfer and high open-circuit voltage
3. High planarity and crystallinity to ensure a high hole mobility
4. Good compatibility with acceptor materials to form stable nanoscale morphology
5. Good processability and high photostability

In early research, poly[2-methoxy-5-(2′-ethylhexyloxy)-p-phenylene vinylene] (MEH-PPV) and its analogue poly[2-methoxy-5-(3′,7′-dimethyloctyloxy)-p-phenylene vinylene] (MDMO-PPV) was used as a n-type donor material in PSCs [9, 10]. Shaheen et al. have reported a bulk heterojunction PSC based on MDMO-PPV and PCBM. By controlling the morphology of phase separation of the blend, an efficiency of 2.5 % was achieved [11]. However, further increase in performance is hindered by the relatively low hole mobility and narrow spectrum availability.

Soluble poly(3-alkylthiophene)s with higher hole mobility, narrower bandgap, and good processability become competent successors [12–15]. Currently, regioregular poly(3-alkylthiophene) is widely used as a polymer donor material in PSCs. The alkyl chain in these polymers has a strong impact on the solubility, crystallinity, and morphology of the polymers. A poly(3-alkylthiophene) with a shorter side chain, like poly(3-butylthiophene), has limited solubility and crystallinity, leading to decreased hole mobility. Longer alkyl chain renders the polymer reduced absorption coefficient because the chromophores, conjugated backbones, are diluted by the alkyl pendants. Among poly(3-alkylthiophene) homologs, poly(3-hexylthiophene) (P3HT) has achieved the highest efficiency. PSCs based on P3HT-PCBM system have realized high energy conversion efficiencies to 4–5 % [14, 15]. P3HT has a bandgap of 1.9 eV and hole mobility of ~0.1 $cm^2\ V^{-1}\ s^{-1}$. Its molecular weight and regularity dictate the performances of PSCs. Regioregular P3HT promises a better crystallinity that is beneficial for hole transport and phase separation. The performance of P3HT-PCBM PSC can be optimized through carefully controlling the morphology of active materials. Thermal annealing has been proved an effective strategy which induces the recrystallization and reduces defects at the interface.

4.1.3.2 Electron Transport Layer

As a part of cathode, the electron transport layer plays an especially important role in enhancing the PCE benefited from the improved charge transport and charge extraction. Moreover, the electron transport layer improves device stability as an oxygen scavenger that can protect against degradation of bulk heterojunction in air.

In most cases, cathode in a PSC is made of low work function metals like calcium, barium, and magnesium to extract electrons from the LUMO of active materials. But these metals are reactive and expensive. Inheriting the success in organic light-emitting diodes (OLEDs), lithium fluoride–aluminum was introduced as cathode in PSCs [16]. The effect of LiF is less understandable. One explanation is the thin layer of LiF may decompose and dope the organic semiconductor and then lower the work function [5, 6].

Inorganic compounds such as titanium oxide (TiO_x), zinc oxide (ZnO), and cesium carbonate (Cs_2CO_3) have been widely investigated for efficient electron extraction in organic photovoltaics [17–20]. Merits of these inorganic compounds include moderate preparation conditions, relatively high electron mobility, environmental stability, and high transparency. TiO_2 and ZnO are the most important materials as electron transport layer in organic photovoltaics. But the high work function (~4.1 eV) of ZnO, the charge extraction barrier between ZnO, and the adjacent photoactive layer only allow moderately efficient PSCs. Introducing self-assembled monolayers or physically absorbed interlayers with strong dipole moments is a conducive approach to modify the work function of the electrode. Moreover, carbon nanotube and graphene with high conductivity, mechanical strength, and thermal stability are considered as promising material for electron transport layer [21, 22].

4.1.3.3 Hole Transport Layer

In the working electrode adjacent with polymer donor materials, hole transport layer requires the materials to have a high work function to match the HOMO level of photoactive layer in order to collect holes and block electrons. High work function PEDOT:PSS (5.0 eV) and metal oxides such as molybdenum oxide (MoO_3), tungsten oxide (WO_3), or vanadium pentoxide (V_2O_5) are effective candidates. PEDOT:PSS and MoO_3 have high transparency and conductivity. PEDOT: PSS film can be easily obtained by spin coating which is suitable for large-scale production, while MoO_3 layer requires thermal evaporation which significantly limited their prospects. However, recently some groups have demonstrated that MoO_3 or WO_3 hole transport layer can also be applied through a facile precursor route from molybdenum(V) ethoxide, tungsten(V) ethoxide, or tungsten(VI) ethoxide [23]. At the same time, carbon nanotubes with high work function and conductivity are also used as the hole transport layer [24, 25].

4.1.3.4 Electrodes

PSCs are commonly thin-film devices fabricated in the sandwich configuration. The two electrodes are usually metallic electrode (such as Al or Ag) and transparent conductive electrode such as indium tin oxide (ITO) glass. Recently, several potential materials based on silver nanowires, carbon nanotubes, and graphene emerged which can extend the PSC to applications pertaining to transparent and flexible devices.

4.1.4 Characterization

Four parameters are used to gauge the performance of a PSC: open-circuit voltage (V_{OC}), short-circuit current density (J_{SC}), fill factor (FF), and PCE (η). They are measured under AM 1.5 G spectrum, which simulates the solar irradiation on the earth surface with an incident angle of 48.2°. The irradiation intensity (the amount of radiant energy received from the sun per unit area per unit time) is normalized to 1,000 W m^{-2}. At standardized condition, the PCE of a PSC is given as

$$\eta = \frac{V_{OC} J_{SC} FF}{I_0} \tag{4.1}$$

where I_0 is the intensity of the incident light, which is 1,000 W m^{-2} at AM 1.5 G. FF is defined as the ratio of the maximum power (P_{max}) and the product of V_{OC} and J_{SC}.:

$$FF = \frac{P_{max}}{V_{OC} J_{SC}} \times 100\ \% \tag{4.2}$$

The maximum power is obtained at the point of *J-V* curve where the product of the photocurrent and photovoltage is maximal.

Another important parameter to characterize the PSC is the incident photon-to-current conversion efficiency (IPCE), also named external quantum efficiency (EQE), which provides information about the monochromatic quantum efficiencies of a solar cell. It is defined to the photocurrent density produced in the external circuit under monochromatic illumination of the cell divided by the photon flux that strikes the cell:

$$IPCE = \frac{J_{SC}(\lambda)}{e\Phi(\lambda)} = 1,240 \frac{J_{SC}(\lambda)[A \bullet cm^{-2}]}{\lambda[nm] I_0(\lambda)[W \bullet cm^{-2}]} \tag{4.3}$$

Practically, IPCE is determined by a sequence of processes: light absorption (A), exciton diffusion (D), charge separation (S), charge transfer (T), and charge collection (C). Therefore, we have

$$IPCE(\lambda) = \eta_A \times \eta_D \times \eta_S \times \eta_T \times \eta_C \tag{4.4}$$

IPCE has a strong impact on the short-circuit current density which will be elucidated in the following discussion.

4.1.4.1 Open-Circuit Voltage

Open-circuit voltage, an energetic index to a PSC, is derived from the energy difference between the HOMO of donor and LUMO of acceptor. It can be empirically expressed as

$$V_{OC} = [|E_{LUMO}(\text{acceptor}) - E_{HOMO}(\text{donor})| - 0.3\ \text{eV}]/e \tag{4.5}$$

The constant number 0.3 eV is the empirical factor for charge separation [26]. On this account, donor polymer with a lower HOMO promises a higher V_{OC}. For example, P3HT has a HOMO of ~4.9 eV, rendering a V_{OC} of 0.6 V [27]. Chemical modification, like introducing fluorene moieties to polymer, can shift the energy level of the donor and increase the open-circuit voltage [28]. The open-circuit voltage is sensitive to the interface between electrode and active materials as well. Tuning the work function of the electrode, as mentioned in last section, can achieve a higher voltage.

4.1.4.2 Short-Circuit Current Density

The kinetic behavior of a PSC is often indicated by the short-circuit current density which involves the processes of charge generation, transfer, and collection. It corresponds to the integration of cell responsivity and incident irradiance [6]. Hence, narrowing the bandgap for a broader spectrum sensitivity is beneficial for a larger photocurrent. A more quantitative expression at ideal circumstance is given as

$$J_{SC} = ne\mu E \tag{4.6}$$

where n is the charge carrier density, μ is the charge carrier mobility, and E is the electric field intensity [5]. The mobility of charge carriers is more related to the morphology of materials than their intrinsic properties, which requires delicate design and preparation. Additionally, varying the acceptor material, for example, displacing C_{60} derivatives with C_{70} derivatives, can also realize an improvement in J_{SC} [6].

In principle, increasing V_{OC} as well as J_{SC} is a powerful strategy towards high efficiency. But technically, it is more complicated. Simultaneously elevating the photovoltage and photocurrent is still a formidable challenge. Raising the HOMO of donor material can narrow the bandgap to improve J_{SC}, but it will compromise the V_{OC} in the meantime.

4.1.4.3 Fill Factor

Fill factor is determined by the number of charge carriers collected by the electrodes, especially when the built-in field is nearing the open-circuit voltage. It is multilaterally affected by charge carrier mobility, nanoscale morphology, compatibility between the materials, series and shunt resistances of device, and charge recombination during transport [5]. The fill factor is less understood and is sensitive to both materials and device design.

4.1.5 Summary

The creation of high-performance PSC is marked as a great achievement in photovoltaic devices. Currently, the highest efficiency of PSC is beyond 11 % and researches towards higher efficiency are still underway. In this section, we have provided a rudimentary introduction of PSCs. The improvement in PSC is a systematic engineering which involves molecular design, material preparation, and device fabrication. A sketchy understanding is listed below:

1. Narrowing the bandgap of polymer donor material is a major approach in molecular engineering for a higher J_{SC}.
2. Lowering the HOMO of the donor material is beneficial for a higher V_{OC}.
3. Higher mobility is essential to photocurrent and fill factor which should be considered in material preparation and morphology control.
4. Introducing bulk heterojunction significantly enlarges the interface between the donor and acceptor, which highlights the importance of nanoscale morphology in a blend.
5. The stability is another problem in PSCs. More attention should be paid into the stability of conjugated polymer and blend under long-term exposure in sunlight and their resistance against oxygen and humidity.

4.2 Overview of Fiber-Shaped Polymer Solar Cell

Nowadays, it is highly desired to develop miniature energy harvesting and storage devices for portable and foldable electronics. In this section, fiber-shaped polymer solar cells (PSCs) that exhibit unique and promising advantages such as lightweight and weaveability are presented. New materials and innovative designs are highlighted for these fiber-shaped energy harvesting devices. Compared with the dye-sensitized solar cells (DSCs) that often require liquid electrolytes, PSCs are all-solid-state devices. Therefore, it is particularly meaningful to develop fiber-shaped PSCs for various electronic fabrics. Similar to the DSCs, both twisted and coaxial structures are developed.

4.2.1 *Fiber-Shaped Polymer Solar Cells in Twisting Structure*

The first fiber-shaped PSC was materialized in 2009, based on two metal wires. The main parts of the PSC, electron transport layer (TiO_x), photoactive layer (P3HT: PCBM), and hole transport layer (PEDOT:PSS), were constructed on a metal wire by dip coating. The other electrode, a silver coated wire, was twisted with the primary electrode to conduct electrons [29]. The two electrodes were encased with a transparent polymer cladding. The twisted structure ensured the electrical contact between the two electrodes, and the resulting PSCs gave an average efficiency of 2.99 %. However, the superficially wrapped secondary electrode would shadow the illuminated area of the primary electrode. It is interesting that the highest PCE, 3.81 % delivered from the fiber-shaped PSC, was obtained provided that the secondary electrode was placed beneath the primary electrode to eliminate the shadowing effect.

Technically, internal stress is inevitably generated when two metal wires are twisted. Therefore, the metal wire-based twisted structure is not mechanically stable and the two electrodes are easily to slip off. To this end, a more flexible fiber electrode such as carbon nanotube (CNT) fiber is preferred as the counter electrode in PSC. For instance, a CNT fiber was twisted with a modified steel with ZnO nanocrystal layer as a hole-blocking layer and P3HT:PCBM as a photoactive layer, the resulting PSC showing an efficiency of 2.11 % (Fig. 4.3) [30]. Displacing silver wire as the counter electrode under the same condition, a much lower efficiency of 0.8 % was obtained, which highlights the importance of the flexible CNT fiber. A flexible electrode is favorable for an intimate and conformal contact with the primary electrode, making the cell insusceptible to deformations.

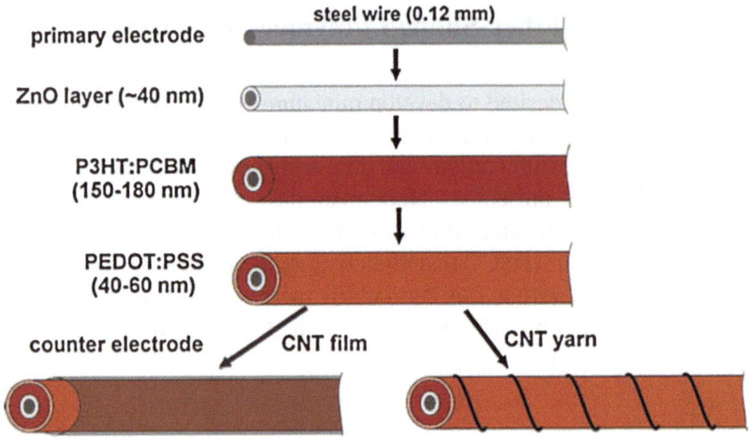

Fig. 4.3 Fabrication of fiber-shaped PSC in twisting and coaxial structures (Reprinted with the permission from Ref. [30]. Copyright 2012 American Chemical Society)

4.2.2 *Fiber-Shaped Polymer Solar Cells in a Coaxial Structure*

Since the charge generation and separation take place at the primary electrode in fiber-shaped PSC, the counter electrode plays an important role in conducting the electrons. For the twisting structure, the low contact area between two fiber electrodes may limit the charge extraction to the external circuit. On this account, a coaxial structure where the counter electrode is wrapped around the primary electrode is conducive for charge collection. Two-dimensional films like CNT sheets and graphene films are preferred as counter electrode materials due to their high surface area, flexibility, and electrical conductivity. Replacing the CNT fiber with single-walled CNT film as counter electrode, the efficiency recorded was 2.31 % (Fig. 4.3) [30]. The major improvement compared with the twisting structure is the fill factor, which increased from 44 to 50 %, indicating an enhanced charge-collection efficiency attributed to the CNT film.

Apart from CNT films, graphene with intrinsic two-dimensional conformation was also applied as counter electrode in coaxial PSC [31]. As presented in Fig. 4.4a, a single-layer graphene sheet was prepared from chemical vapor deposition (CVD) on a copper substrate. Then the graphene sheet was introduced with gold nanoparticles to enhance its conductivity and then transferred to wrap around the primary electrode. Compared with CNT films, the single-layer graphene had a higher optical transmission (95 %). The high flexibility of the single-layer graphene

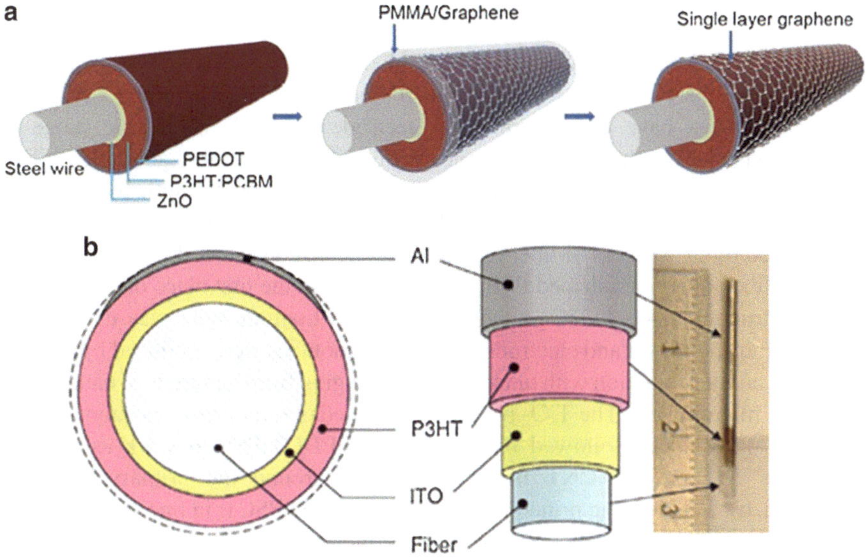

Fig. 4.4 **a** Preparation of a coaxial fiber-shaped PSC based on graphene (Reproduced from Ref. [31] by permission of The Royal Society of Chemistry). **b** A coaxial fiber-shaped PSC based on an optical fiber (Reprinted with permission from Ref. [32]. Copyright 2007 by AIP Publishing LLC)

also rendered a conformal contact with the electrode. An efficiency of 2.53 % was obtained with V_{oc} of 0.570 V, J_{sc} of 8.14 mA cm^{-2}, and fill factor of 54.5 %. Moreover, the intact graphene layer served as physical barrier to protect the enclosed active materials from oxygen and humidity. As demonstrated, the efficiency degradation was less than 5 % after several days' exposure in air.

The coaxial structure can also realize a more integrated device similar with the planar PSC. For example, a PSC with a core-sheath structure is presented in Fig. 4.4b, where the electron transport layer, photoactive layer, and hole transport layer are sheathed on the ITO-modified optical fiber [32]. Optical fibers are used as substrates in the core-sheath architecture as the incident light can be conducted into the photoactive layer along the coaxial direction. The conductive ITO layer and photoactive layer of P3HT:PCBM were sequentially coated on the optical fiber by a dip-coating process, and LiF and Al film was then deposited on the outer surface by thermal evaporation method. Compared with the planar counterpart, the reflective and transmissive losses can be minimized and an efficiency of 1.1 % was obtained. Besides the optical fiber, modified plastic fibers such as polyimide-coated silica fiber and polypropylene fiber were also used as the fiber substrate in polymer solar cells, but the coated conductive metal layers are required to be less than 20 nm in thickness for the transmission of the incident light [33].

4.3 Fiber-Shaped Polymer Solar Cells Based on Carbon Nanotubes

In this section, we present a new family of fiber-shaped PSCs based on the aligned carbon nanotube (CNT) materials. Aligned CNT materials, including CNT fiber and CNT sheet, have been studied as electrode materials in a wide range of applications due to their combined properties of light weight, good flexibility, high electrical conductivity (10^2–10^3 S cm^{-1}), mechanical strength (~600 Mpa), and electrocatalytic activity. They also show their advantages in fiber-shaped PSCs.

Inspired by the success of fiber-shaped dye-sensitized solar cells (Chap. 3) based on CNT fibers, the fiber-shaped PSC was developed in the first place on the basis of similar structure [34]. A TiO_2 nanotube–modified titanium wire was used as the electron transport layer and electrode. The TiO_2 nanotubes were produced by electrochemical anodic oxidation with tunable lengths ranging from hundreds of nanometers to tens of micrometers. The TiO_2 nanotube-modified titanium wire was immersed in P3HT:PC_{70}BM blend followed by dip-coating PEDOT:PSS to work as primary electrode. At last, the CNT fiber was twined around the primary electrode (Fig. 4.5a, b). The working principle of this fiber-shaped PSC is in line with its planar counterparts. P3HT molecules absorb incident photons and generate excitons, which then disassociate at the interface with PC_{70}BM and also with TiO_2. The TiO_2 here also acts as the electron acceptor since its conducting band (4.2 eV) is situated beneath the LUMO level of P3HT. The electrons travel through PC_{70}BM or TiO_2 and extracted by

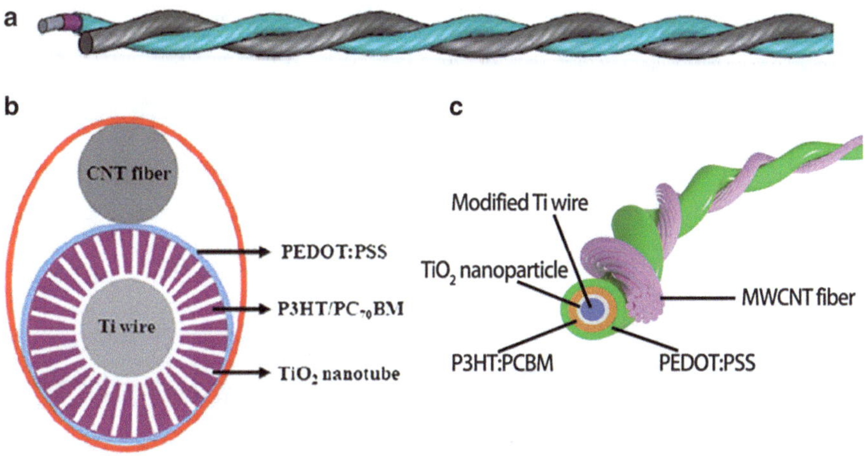

Fig. 4.5 **a** Schematic illustration of fiber-shaped PSCs based on CNT fiber. **b** and **c** PSCs with photoanodes constructed on Ti wire modified with TiO_2 nanotubes (**b**) and TiO_2 nanoparticles (**c**). (**a** and **b** Reproduced from Ref. [34] by permission of The Royal Society of Chemistry **c** Reproduced from Ref. [35] by permission of John Wiley & Sons Ltd)

the Ti wire while the holes migrate through PEDOT:PSS and collected by CNT fiber. The value of photovoltaic parameters recorded are $V_{oc} = 0.42$ V, $J_{sc} = 0.98$ mA cm^{-2}, and FF = 36 %, which gave an efficiency of 0.15 %.

The low PCE was attributed to the ineffective contact between the electrode and photoactive layer which was not favorable for charge separations and transports. To this end, a thin layer of TiO_2 nanoparticles was incorporated between the fiber substrate and photoactive layer (Fig. 4.5c) [35]. The TiO_2 nanoparticle has shown its advantages in dye-sensitized solar cells, and it is also found conducive in PSC. The PCE with TiO_2 nanoparticle introduced was increased by 36 %. The nanocrystalline semiconductor oxide layer played a crucial role in providing pathways for charge transport. In most cases, a smooth and flat surface is favorable for a thin and uniform P3HT:PCBM and PEDOT:PSS layers. However, the uniform surface of Ti wire is disrupted by the grown TiO_2 nanotubes, which is remedied by introducing TiO_2 nanoparticles. Moreover, the nanoparticle layer offers a large surface area for polymer, decreasing the electrical resistance for charge transport and enhancing the light scattering. Note that the outer polymer layer may slightly decrease the light access to the nanoparticle layer. Here this effect was not obvious as the ultrathin polymer layers were transparent with high optical transmittances.

As the electron transport layer, the size of TiO_2 nanotubes including length and diameter has a strong impact on the performance of PSCs. The length of TiO_2 nanotubes on the Ti wire can be controlled by varying the growth time during the electrochemical anodization, ranging from 650 nm to 4 μm. It was found that TiO_2 nanotubes with a length of ~1.8 μm produced the highest PCE (1.01 %). TiO_2 nanoparticles were incorporated by chemical treatment in titanium tetrachloride ($TiCl_4$). Figure 4.6 displays the morphology of the TiO_2 nanotubes before and after

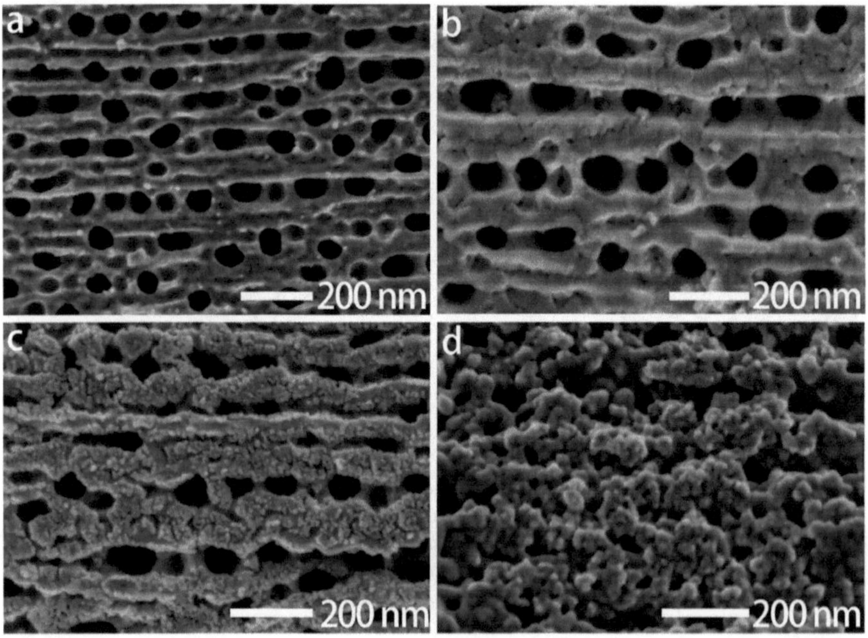

Fig. 4.6 SEM images of TiO_2 nanotubes without and with treatment in $TiCl_4$ for different time. **a** Without treatment. **b** 15 min. **c** 30 min. **d** 60 min (Reproduced from Ref. [35] by permission of John Wiley & Sons Ltd)

treating for different time. TiO_2 nanoparticles had a uniform size of ~20 nm and were apt to aggregate into large agglomerates when treated for 60 min.

The mechanism of the fiber-shaped PSC is schematically shown in Fig. 4.7a. The layer of TiO_2 nanoparticles exerted a significant effect on fiber-shaped PSCs. Figure 4.7b shows *J*-*V* curves of PSCs before and after being treated by a $TiCl_4$ solution. The effective illuminated area was calculated by multiplying the length and diameter of the primary electrode including the thickness of Ti wire, TiO_2 nanotubes, and photoactive materials. The PCE was beneath 1.31 % without the treatment with V_{oc} of 0.50 V, J_{sc} of 6.48 mA cm^{-2}, and fill factor of 41 %. The efficiencies were increased to 1.60 % and 1.78 % after treating for 15 and 30 min.

The agglomeration of TiO_2 is detrimental to PSC as verified by its decreased efficiency of 1.38 % with 60 min treatment. The evolution of efficiency with treating time is backed up by IPCE curves illustrated in Fig. 4.7c. The diameters of CNT fibers were also found to largely affect the fiber-shaped PSCs (Fig. 4.7d). The PCE reached its peak when the diameter of CNT was 32 μm, which was the compromise of electrical conductivity and illuminated area. A thinner fiber has a higher electric resistance leading to a lower current density and fill factor, whereas a thicker fiber shadows more incident light and also decreases the current density.

A wide variety of metal wires have been studied as surrogates of CNT fiber but with inferior performances. For instance, a PSC with a silver wire as the counter

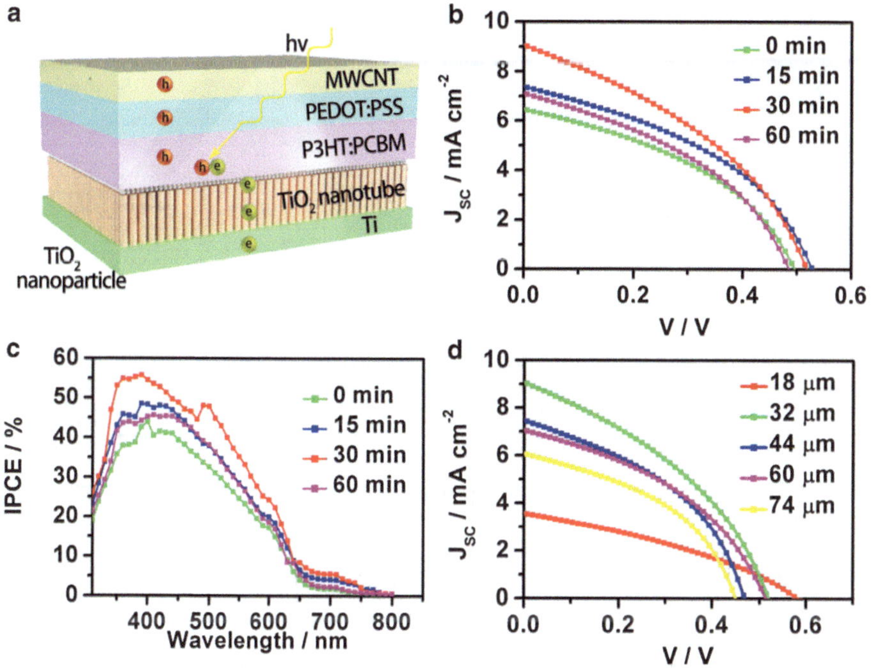

Fig. 4.7 **a** Schematic illustration to the mechanism of a PSC. **b**, **c**. J-V and IPCE curves of PSCs after treated by $TiCl_4$ for different times, respectively. **d** *J-V* curves of PSCs with CNT fibers in increasing diameters (Reproduced from ref. [35] by permission of John Wiley & Sons Ltd)

electrode delivered an efficiency of 0.48 %. The advantage of CNT fiber is represented by two aspects. First, the work function of CNT fiber (4.8–5.1) eV is more suitable for the hole collection compared with that of silver wire. Second, the metal wires were not sufficiently flexible for fiber-shaped PSCs to afford high efficiencies. As shown in Fig. 4.8a–d, the two electrodes were not in intimate contact. The gap appeared is unfavorable for an efficient charge transfer, which becomes worse under bending. The PCE degraded from 0.48 to 0.14 % after bending by 50 cycles.

Beneficial from the flexible electrodes, the fiber-shaped PSCs were flexible and could be easily deformed into various structures. After enduring bending for 1,000 times, the PSC survived with ~85 % of efficiency retained (Fig. 4.8e). The fiber-shaped PSC also exhibited a decent stability. For instance, the efficiency retained over 70 % when exposed in air after 16 days (Fig. 4.8f).

The as-prepared fiber-shaped PSC can be easily woven into various flexible structures such as fabrics without the necessity for sealing that is required for fiber-shaped dye-sensitized solar cells (Fig. 4.9a, b). Weaving into fabric is an effective strategy to integrate PSC units into a tandem device. The output voltages are linearly increased with the increasing number of fiber-shaped PSC units (Fig. 4.9c). As expected, the PSC fabric is highly flexible, and the energy

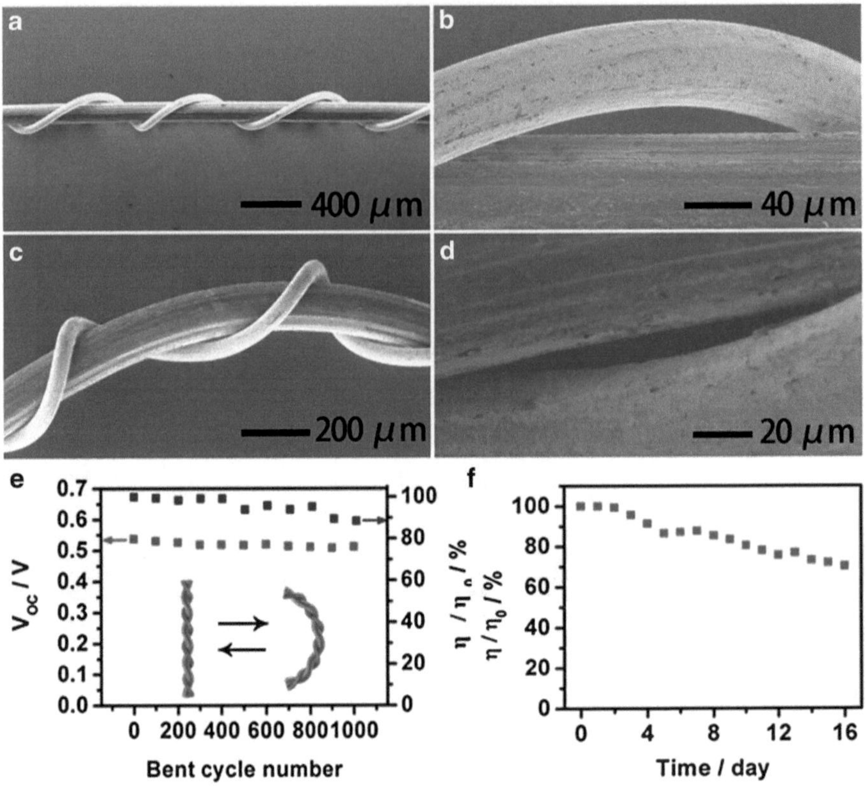

Fig. 4.8 **a–d** SEM images of a fiber-shaped PSC based on the silver wire. **a** and **b** Natural state. **c** and **d** Bent state. **e** Dependence of PCE and voltage on bent cycle number. η_0 and η correspond to the efficiencies before and after bending, respectively. **f** Evolution of PCE with time upon exposure in air. η_0 and η correspond to the PSC before and after exposed in air, respectively (Reproduced from ref. [35] by permission of John Wiley & Sons Ltd)

conversion efficiencies retained by ~80 % after bending for 1,000 cycles (Fig. 4.9d). Interestingly, the performances of PSC fabrics are independent of the incident angle, e.g., the energy conversion efficiencies were varied in less than 20 % in the different incident angles (Fig. 4.9e). Displacing CNT fiber with CNT sheet, we can obtain a coaxial PSC with maximal efficiency of 1.01 %.

4.4 Perspective

Achievements in fiber-shaped PSC are encouraging. But it is too early to acclaim a triumph. There are many challenges in fiber-shaped PSCs. Their efficiencies are still much lower than the conventional planar PSCs. In particular, the lengths of

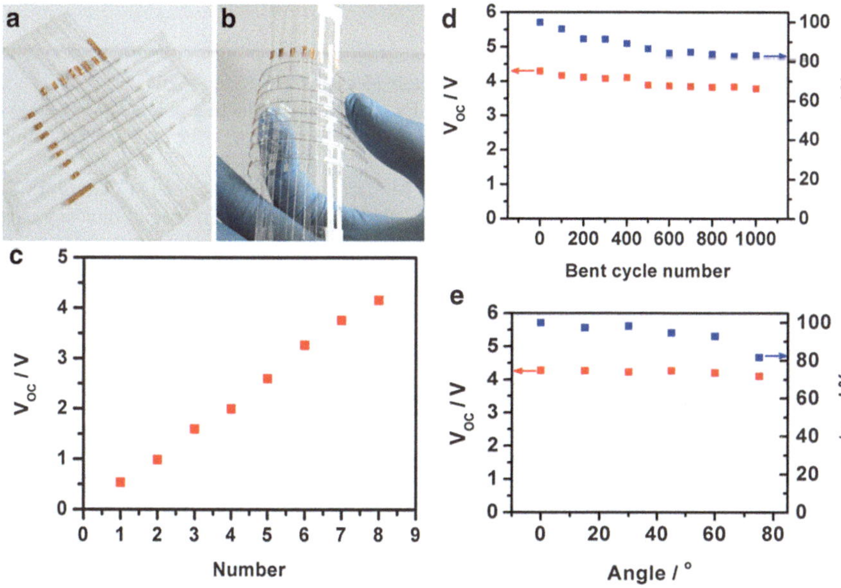

Fig. 4.9 **a** and **b** A PSC fabric in natural and deformed state, respectively. **c** Dependence of voltage on the number of fiber-shaped PSC units in a tandem PSC fabric. **d** Dependence of PCE and voltage on bent cycle number in a PSC fabric. η_0 and η correspond to the efficiencies before and after bending, respectively. **e** Dependence of PCE and voltage on the angle between the solar cell textile and substrate with the light source at the top. η_0 and η correspond to the 0 and the other angle, respectively (Reproduced from ref. [35] by permission of John Wiley & Sons Ltd)

available fiber-shaped PSCs ranged from several to tens of centimeters. Increasing the cell length to an applicable level, the efficiency will slump as the upsurge of electrical resistances. Currently it is still difficult to realize a balance between high efficiencies and acceptable lengths. To this end, the fiber electrodes with high electrical conductivities are critically important to develop usable fiber-shaped solar cells.

Technical challenges in fabricating fiber-shaped PSCs still exist. For a core-sheath structure, it is difficult to coat uniform and thin photoactive and electrode layers on a fiber substrate, particularly, at a large scale based on the current fabrication technology. For a twisting architecture, it is difficult to wind two fiber electrodes in a well-tunable and repeatable process. It is also difficult to make two fiber electrodes closely and stably intertwined during the use. In addition, a promising advantage of fiber-shaped PSCs lies in that they can be woven into clothes or other flexible structures. However, it remains unclear how the electrodes are connected to collect the electric power for so many wire devices. To the best of our knowledge, no attempts have been made to address this concern and develop technologies for such organizations.

References

1. Tang CW (1979) Multilayer organic photovoltaic elements. US Patent 4,164,431
2. Sariciftci NS, Smilowitz L, Heeger AJ, Wudl F (1993) Semiconducting polymers (as donors) and buckminsterfullerene (as acceptor): photoinduced electron transfer and heterojunction devices. Synth Met 59(3):333–352
3. You J, Dou L, Yoshimura K, Kato T, Ohya K, Moriarty T, Emery K, Chen C-C, Gao J, Li G, Yang Y (2013) A polymer tandem solar cell with 10.6 % power conversion efficiency. Nat Commun 4:1446
4. Chen C-C, Chang W-H, Yoshimura K, Ohya K, You J, Gao J, Hong Z, Yang Y (2014) An efficient triple-junction polymer solar cell having a power conversion efficiency exceeding 11 %. Adv Mater 26(32):5670–5677
5. Günes S, Neugebauer H, Sariciftci NS (2007) Conjugated polymer-based organic solar cells. Chem Rev 107(4):1324–1338
6. Li G, Zhu R, Yang Y (2012) Polymer solar cells. Nat Photonics 6(3):153–161
7. Kim JY, Lee K, Coates NE, Moses D, Nguyen T-Q, Dante M, Heeger AJ (2007) Efficient tandem polymer solar cells fabricated by all-solution processing. Science 317(5835):222–225
8. Sariciftci NS, Smilowitz L, Heeger AJ, Wudl F (1992) Photoinduced electron transfer from a conducting polymer to buckminsterfullerene. Science 258(5087):1474–1476
9. Yu G, Gao J, Hummelen JC, Wudl F, Heeger AJ (1995) Polymer photovoltaic cells: enhanced efficiencies via a network of internal donor-acceptor heterojunctions. Science 270 (5243):1789–1791
10. Wudl F, Srdanov G (1993) Conducting polymer formed of poly (2-methoxy, 5-(2′-ethyl-hexyloxy)-p-phenylenevinylene). US Patent 5,189,136
11. Shaheen SE, Brabec CJ, Sariciftci NS, Padinger F, Fromherz T, Hummelen JC (2001) 2.5 % efficient organic plastic solar cells. Appl Phys Lett 78(6):841–843
12. Padinger F, Rittberger RS, Sariciftci NS (2003) Effects of postproduction treatment on plastic solar cells. Adv Funct Mater 13(1):85–88
13. Bao Z, Dodabalapur A, Lovinger AJ (1996) Soluble and processable regioregular poly (3-hexylthiophene) for thin film field-effect transistor applications with high mobility. Appl Phys Lett 69(26):4108–4110
14. Li G, Shrotriya V, Huang J, Yao Y, Moriarty T, Emery K, Yang Y (2005) High-efficiency solution processable polymer photovoltaic cells by self-organization of polymer blends. Nat Mater 4(11):864–868
15. Ma W, Yang C, Gong X, Lee K, Heeger AJ (2005) Thermally stable, efficient polymer solar cells with nanoscale control of the interpenetrating network morphology. Adv Funct Mater 15 (10):1617–1622
16. Brabec CJ, Shaheen SE, Winder C, Sariciftci NS, Denk P (2002) Effect of LiF/metal electrodes on the performance of plastic solar cells. Appl Phys Lett 80(7):1288–1290
17. Za T, Zhang W, Zhang Z, Qian D, Huang Y, Hou J, Li Y (2012) High-performance inverted polymer solar cells with solution-processed titanium chelate as electron-collecting layer on ITO electrode. Adv Mater 24(11):1476–1481
18. Kuwabara T, Nakayama T, Uozumi K, Yamaguchi T, Takahashi K (2008) Highly durable inverted-type organic solar cell using amorphous titanium oxide as electron collection electrode inserted between ITO and organic layer. Sol Energy Mater Sol Cells 92(11):1476–1482
19. Sun Y, Seo JH, Takacs CJ, Seifter J, Heeger AJ (2011) Inverted polymer solar cells integrated with a low-temperature-annealed sol-gel-derived ZnO film as an electron transport layer. Adv Mater 23(14):1679–1683
20. Colsmann A, Reinhard M, Kwon T-H, Kayser C, Nickel F, Czolk J, Lemmer U, Clark N, Jasieniak J, Holmes AB, Jones D (2012) Inverted semi-transparent organic solar cells with spray coated, surfactant free polymer top-electrodes. Sol Energy Mater Sol Cells 98:118–123

21. Lu L, Xu T, Chen W, Lee JM, Luo Z, Jung IH, Park HI, Kim SO, Yu L (2013) The role of N-doped multiwall carbon nanotubes in achieving highly efficient polymer bulk heterojunction solar cells. Nano Lett 13(6):2365–2369
22. Jung YS, Hwang Y-H, Javey A, Pyo M (2011) PCBM-grafted MWNT for enhanced electron transport in polymer solar cells. J Electrochem Soc 158(3):A237–A240
23. Höfle S, Bruns M, Strässle S, Feldmann C, Lemmer U, Colsmann A (2013) Tungsten oxide buffer layers fabricated in an inert sol-gel process at room-temperature for blue organic light-emitting diodes. Adv Mater 25(30):4113–4116
24. Bindl DJ, Ferguson AJ, Wu M-Y, Kopidakis N, Blackburn JL, Arnold MS (2013) Free carrier generation and recombination in polymer wrapped semiconducting carbon nanotube films and heterojunctions. J Phys Chem Lett 4(21):3550–3559
25. Stylianakis MM, Kymakis E (2012) Efficiency enhancement of organic photovoltaics by addition of carbon nanotubes into both active and hole transport layer. Appl Phys Lett 100 (9):093301
26. Scharber MC, Mühlbacher D, Koppe M, Denk P, Waldauf C, Heeger AJ, Brabec CJ (2006) Design rules for donors in bulk-heterojunction solar cells—towards 10 % energy-conversion efficiency. Adv Mater 18(6):789–794
27. Yun J-J, Jung H-S, Kim S-H, Han E-M, Vaithianathan V, Jenekhe SA (2005) Chlorophyll-layer-inserted poly(3-hexyl-thiophene) solar cell having a high light-to-current conversion efficiency up to 1.48 %. Appl Phys Lett 87(12):123102
28. Zhou Q, Hou Q, Zheng L, Deng X, Yu G, Cao Y (2004) Fluorene-based low band-gap copolymers for high performance photovoltaic devices. Appl Phys Lett 84(10):1653–1655
29. Lee MR, Eckert RD, Forberich K, Dennler G, Brabec CJ, Gaudiana RA (2009) Solar power wires based on organic photovoltaic materials. Science 324(5924):232–235
30. Liu D, Zhao M, Li Y, Bian Z, Zhang L, Shang Y, Xia X, Zhang S, Yun D, Liu Z, Cao A, Huang C (2012) Solid-state, polymer-based fiber solar cells with carbon nanotube electrodes. ACS Nano 6(12):11027–11034
31. Liu D, Li Y, Zhao S, Cao A, Zhang C, Liu Z, Bian Z, Liu Z, Huang C (2013) Single-layer graphene sheets as counter electrodes for fiber-shaped polymer solar cells. RSC Adv 3 (33):13720–13727
32. Liu J, Namboothiry MAG, Carroll DL (2007) Optical geometries for fiber-based organic photovoltaics. Appl Phys Lett 90(13):133515
33. Bedeloglu A, Demir A, Bozkurt Y, Sariciftci NS (2010) A photovoltaic fiber design for smart textiles. Text Res J 80(11):1065–1074
34. Chen T, Qiu L, Li H, Peng H (2012) Polymer photovoltaic wires based on aligned carbon nanotube fibers. J Mater Chem 22(44):23655–23658
35. Zhang Z, Yang Z, Wu Z, Guan G, Pan S, Zhang Y, Li H, Deng J, Sun B, Peng H (2014) Weaving efficient polymer solar cell wires into flexible power textiles. Adv Energy Mater 4(11):1301750

Chapter 5
Fiber-Shaped Perovskite Solar Cell

Abstract This chapter refers to the recent breakthrough of perovskite solar cell as the pioneer in the next-generation photovoltaics. The working mechanism explains the high performance; various structure including mesoscopic structure, meso-superstructure, and thin-film structure are compared; and extensive materials are carefully and systemically discussed. Then the flexible photovoltaics were demonstrated in detail, evolving to the fiber-shaped perovskite solar cell. Furthermore, the fiber-shaped perovskite solar cell realizes stretchability through delicate structure design. Finally, the perspective for the further development tendency was presented.

5.1 Overview of Perovskite Solar Cell

Perovskite, a calcium titanium oxide mineral ($CaTiO_3$) named after Russian mineralogist L. A. Perovski (1792–1856), has lent its name to a classic of organic–inorganic hybrid compound with a chemical formula of ABX_3, where A and B are two cations and X is an anion bonding with them. In a typical unit cell shown in Fig. 5.1, eight A cations are located at corners of the cube, while a B cation is installed at the body center coordinated by six X anions at the face centers. In photovoltaic devices, a specific type of perovskite is preferred where A is organic, generally methylammonium ($CH_3NH_3^+$), B is usually lead or tin, and X is a halogen like bromine or iodine. In this case, ammonium trihalogen plumbates like $CH_3NH_3PbI_3$ often come across in perovskite solar cells.

The photovoltaic application of perovskites is inspired by their early services in thin-film transistors and light-emitting diodes (LEDs) [1]. Their exercise in solar cells started in 2006, when Miyasaka and coworkers employed a perovskite, $CH_3NH_3PbBr_3$, as a visible sensitizer for TiO_2 in a sensitized solar cell and achieved a power conversion efficiency (PCE) of 2.2 % [2]. Afterwards, replacing the halogen anions with iodine, the PCE increased to 3.8 % benefited from a broader spectral sensitivity [3]. Park et al. fabricated a 6.5 % efficient quantum-dot-sensitized solar cell using 2–3 nm perovskite $CH_3NH_3PbI_3$ nanocrystals [4]. However, the perovskite's nonresistance to organic solvent makes the solar cell unstable. On this account, getting rid of the liquid electrolyte, i.e., introducing a solid-state hole conductor, was underway.

H. Peng, *Fiber-Shaped Energy Harvesting and Storage Devices*, Nanostructure Science and Technology, DOI 10.1007/978-3-662-45744-3_5

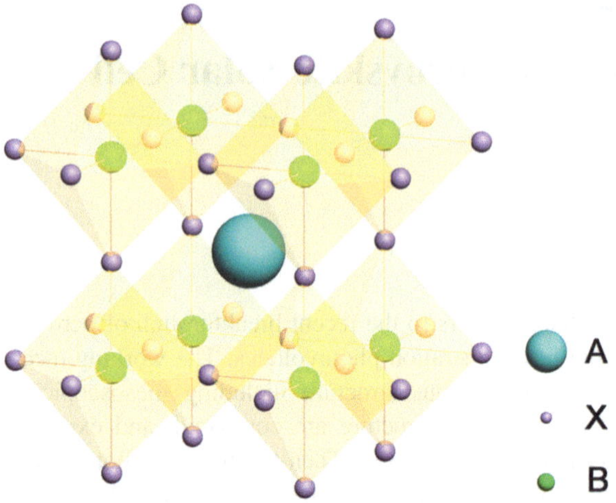

Fig. 5.1 Crystal structure of halide perovskites with the generic chemical formula ABX_3

A polypyrrole-based conducting polymer material has been used as hole transport layer instead of problematic redox electrolyte [5]. As the conducting polymer did not fully infiltrate into the TiO_2 mesoporous structure, the results were not encouraging. 2,2′,7,7′-Tetrakis(*N*,*N*-di-p-methoxyphenylamine)-9,9′-spirobifluorene (spiro-MeOTAD), a solid-state hole conductor, has shown its efficacy to form heterojunction with TiO_2 in dye-sensitized solar cells (DSC) [6]. Naturally, the benefits of spiro-MeOTAD were then transferred to perovskite-sensitized solar cells though a solution-based incorporation and a much improved stability, and PCE of 9.7 % was achieved [7]. In abovementioned systems born out of DSCs, the perovskites performed as a sensitizer to harvest light and inject the excited electrons to the conducting band of the semiconductor crystals. The introduction of solid-state hole conductor, as a successor to the liquid electrolyte which requires an extra energy to drive the redox shuttles and hence imposes a penalty in photovoltage, leapt the hurdle of the solvent incompatibility of the perovskites [8]. Until mid-2012, Snaith and colleagues have reported revolution in paradigm that made the perovskite solar cell branch out from the solid-state DSC as a burgeoning and powerful rival to conventional solar cells [9]. The breakthrough lies in replacing the semiconductive TiO_2 with insulated Al_2O_3 as mesoporous scaffold for extreme thin perovskite coatings. Spiro-MeOTAD was infiltrated into and capped above the scaffold. In this case that Al_2O_3 did not act as an n-type semiconductor, the perovskite took charge of light harvest as well as electron conduction, and the solar cell was not a sensitized type but a two-component hybrid solar cell. The high PCE of 10.7 % with high photovoltage (0.98 V) and photocurrent (17.8 mA cm^{-2}) predicted the superiority of perovskite solar cell that benefits from the broad spectral sensitivity and low bandgap. The perovskite solar cell has vigorously progressed afterwards, engendering diversified device configuration and materials as well as constantly refreshed efficiency. Through delicate manipulation of interface and perovskite morphology, to date, the highest PCE of 19.3 % has been recorded [10].

Inheriting high efficiency as well as solid state, yet it is arbitrary to assert that the perovskite solar cell outperforms its ancestors like dye-sensitized solar cell and polymer solar cell (PSC). Several issues like stability and toxicity are still lingering around, and pitfalls derived from hysteresis effects stir up the debate on the accuracy of the reported efficiency values [11]. But as DSCs are constrained by liquid electrolytes while PSCs are wrestling with their lower efficiencies, the perovskite solar cell, indeed, edges ahead. It is often the case that attempts to make a new type of device deformable and flexible for wearable applications come after its emergence. Several prototypes of flexible perovskite solar cells sprung up recently including the state-of-the-art fiber-shaped cells which will be presented in this chapter. First, it starts with the working mechanism of perovskite solar cell.

5.1.1 Working Mechanism

Many apparently different types of solar cells actually share two common processes during work: light absorption to create electron-hole pairs and charge separation to generate photovoltage and photocurrent. The light absorbers, in general, are photosensitive semiconductors, which classify the solar cells into different categories: sensitized solar cell, polymer solar cell, and, here, perovskite solar cell. The charge separation, which takes place at a contact interphase that is selective for electrons or holes, dictates the energy available inside the device. In a traditional p-i-n solar cell, p-type and n-type semiconductors are selective for holes and electrons, respectively, while in DSC and PSC, the selection is conducted by different electron and hole conductors, and likewise in perovskite.

Particularly in perovskite solar cells, the light is harvested by the perovskite like $CH_3NH_3PbI_3$ that has a bandgap of 1.55 eV, lower than the semiconducting polymers (~2 eV) and inorganic semiconductors (~3.2 eV), thus promising a broad spectral sensitivity with onset of 800 nm. The strong light absorption of the perovskite enables an outstanding performance of the solar cell since it moderates the requirement for film thickness and facilitates the charge collection. The photo-induced excitons after light absorption have small binding energy (30–75 meV) that suffices for thermal dissociation into free carriers (electrons and holes). It has been proved that the excited perovskite is mainly populated by free carriers rather than their associated pairs [12]. Then, the free carriers transport through the electron conductors and hole conductors, respectively, and get collected at electrodes.

As illustrated in Fig. 5.2, there are seven processes involved in the illuminated perovskite solar cell: photoexcitation (1), electron transfer to ETM (2), and hole transfer to HTM (3); (4) to (7) represent the unwanted recombination process including the reassociation of free carriers (4), back transfer from ETM and HTM to perovskite (5) and (6), and local short circuit at the absent of perovskite (7). For perovskite solar cell, one of their advantages over DSC and PSC is the long

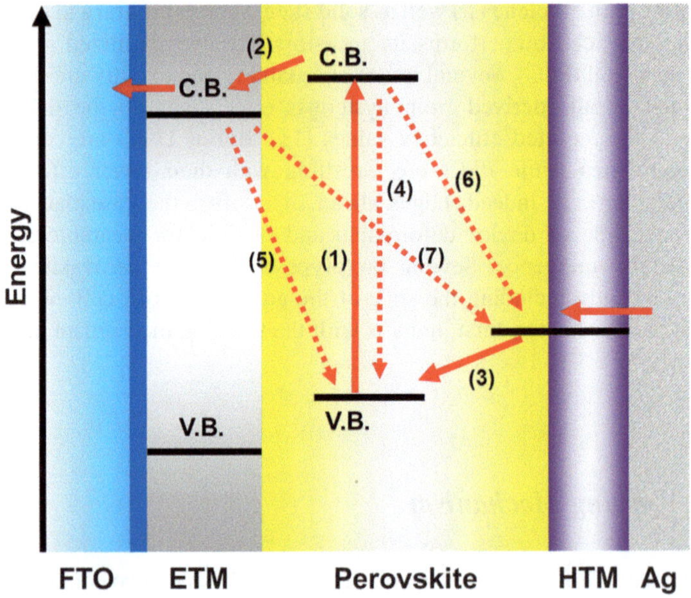

Fig. 5.2 Schematic diagram of energy levels and electron transfer process in an HTM/perovskite/ETM cell

carrier-diffusion length (10^2–10^3 nm) which contributes to a low non-radiative recombination [11].

Another striking advantage is high open-circuit voltage (V_{OC}) generated under illumination. The open-circuit voltage indicates the maximum energy available from absorbed light which is delimited by the bandgap of the perovskite. Normally, the obtained V_{OC} is lower than the bandgap potential (Eg/q). For perovskite solar cells, their V_{OC} ranged from 0.9 to 1.15 V, giving loss-in-potentials about 0.4–0.65 eV, which is better than ~0.75 eV loss-in-potential of the best DSC and ~0.8 eV of PSC [13, 14].

5.1.2 *Structure*

As we have discussed, perovskite solar cell is derived from DSC using perovskites as sensitizers in substitution for organic dyes. In the first place, perovskite solar cell shares the similar sandwich structure with solar cells sensitized by organic dyes and inorganic quantum dots. However, the instability of perovskite in liquid electrolyte motivated the displacement with solid-state electrolytes and HTMs, resulting in a solid-state mesoscopic structure. In light of the ambipolar transport property that perovskite can serve as electron conductor as well as hole conductor, n-type semiconductor (like TiO_2) layer can be displaced with inert mesoporous scaffold

(Al_2O_3) which evolved into meso-superstructured perovskite solar cells. The perovskite as hole conductor, a nanocomposite solar cell where TiO_2 scaffold is infiltrated and capped with perovskite absorber, emerged. Furthermore, the TiO_2 scaffold can be exempted leading to a thin-film p-i-n perovskite solar cell akin to planar heterojunction structure in polymer solar cell. In this section, the structures of mesoscopic solar cell, meso-superstructured solar cell, and thin-film perovskite solar cell are introduced in detail (Fig. 5.3).

5.1.2.1 Mesoscopic Structure

The mesoscopic structure stemmed from extremely thin absorbers cell. It employs broad bandgap mesoporous semiconductors to extract and transport electrons [13]. Perovskites are fully infiltrated into mesoporous semiconductor by solution or vapor deposition [15], and in some cases, they form a capping layer when exceed the absorptive capacity of mesoporous materials. Since perovskite can transport holes, it is different from previous structure that perovskite acts as a quantum dot sensitizer, dispersed on the semiconductor surface. The layered structure is often built up on a transparent conducting tin oxide layer (TCO) deposited with a compact TiO_2 layer with the sputtered Ag as the back electrode. The mesoscopic structure allows a thin perovskite/TiO_2 film (200–300 nm) without sacrificing the photocurrent. As the perovskite is an effective hole conductor, the mesoscopic structure can deliver a high PCE of 12 % [16].

5.1.2.2 Meso-superstructure

The name of meso-superstructure is given by Snaith and colleagues. The emergence of meso-superstructure is marked as a branching point from the previous sensitized analogues. The meso-superstructure is characterized by the nonconductive mesoporous oxide, like Al_2O_3 and ZrO_2, as scaffold instead of the traditional semiconductors, which indicated the perovskite can conduct electrons individually [10, 11]. The conducting band of Al_2O_3 and ZrO_2 is higher than perovskite, impeding the pathway for electron distraction [9]. However, owing to the rapid charge transport and more than 1 μm electron diffusion length of perovskite [17], meso-superstructured solar cells are rational. Impressively, it has been reported that the Al_2O_3-based solar cell exhibited higher open-circuit voltage and faster charge collection [9]. Furthermore, the combination of n-type semiconductor and inert oxide also presented excellent performance and impressive stability.

5.1.2.3 Thin-Film Structure

Thin-film structure, akin to p-i-n heterojunction structure, is a more advanced structure by removing the scaffold materials. The key for achieving high PCE is

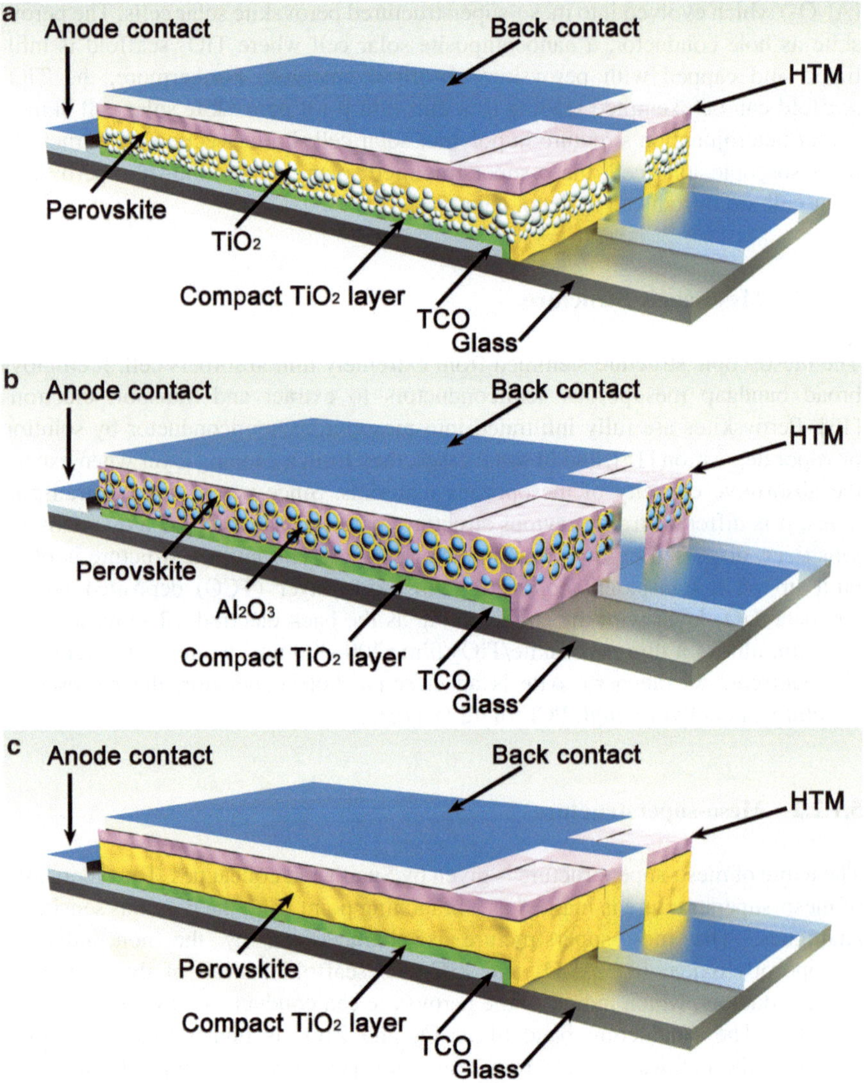

Fig. 5.3 **a** Mesoscopic structure perovskite solar cell where the mesoscopic TiO_2 scaffold is infiltrated by the perovskite. The perovskite assumes the dual role of light absorber and hole conductor. **b** Meso-superstructured perovskite solar cell employing a film of Al_2O_3 nanocrystals covered with a conformal overlayer of perovskite. The latter acts as a light harvester as well as an electron conductor. **c** Cross section of a thin-film structure solar cell lacking the TiO_2 mesoporous scaffold

the uniform thin-film between adequate HTM and ETM. Some researchers have optimized the film morphology through varying solution-cast processing conditions or utilizing vapor deposition. Particularly, by tuning bandgap, this structure can be inverted, where TCO collects holes (generally collecting electrons).

5.1.3 Material

The perovskite solar cell progresses in parallel with the upgrade of materials and deep understanding of working mechanism. In this section, we will focus on the materials of perovskite solar cells, including perovskites, electron, and hole transport materials.

5.1.3.1 Perovskite

Currently, the hybrid organic–inorganic lead halide perovskite takes the dominant position for high-efficiency solar cells which have several extraordinary properties:

1. Various fabrication processes leading to high-quality crystals
2. Adequate bandgap for broad spectrum absorption
3. Ambipolar transport ability enabling it to serve as electron conductors and hole conductors
4. Long diffusion distance to ensure low recombination

At present, $CH_3NH_3PbI_3$ is the most popular perovskite material, but some other perovskite materials, like mixed halide perovskite ($CH_3NH_3PbI_{3-x}Cl_x$), which is more stable in the air, are attracting the increasing attention. Replacing methylammonium cation, the bandgap can be carefully tuned [18]. On the other hand, aiming at avoiding the toxicity of lead element, investigation of lead-free perovskites is necessary. The tin analogues, such as $CH_3NH_3SnI_3$ and $CsSnI_3$ [19, 20], are promising.

5.1.3.2 Electron Transport Materials

ETM is used to assist electron transport and block hole diffusion. Currently, ETM is a necessity in perovskite photovoltaics. In absence of ETM, the electron-hole pairs can hardly disassociate and the recombination prevails.

Compact TiO_2 layer is universal in almost all structures. With appropriate energy levels, it effectively blocks the hole transport, showing a donor density of $\sim 10^{18}$ cm^{-3} [9]. In addition, the modification process, such as yttrium doping, can further increase conductivity (2×10^{-5} S cm^{-1}), due to higher carrier concentration [10]. For mesoscopic structure, porous TiO_2 materials, such as crystals, nanofibers, and nanotubes, reside on the compact layer. Except compact TiO_2 blocking layer, some other materials such as titanium oxide (TiO_x), zinc oxide (ZnO), TiO_2–graphene composites, and fullerene are used.

5.1.3.3 Hole Transport Materials

From the iodide/triiodide liquid electrolyte to spiro-MeOTAD and to HTM free, the evolution of HTM gives impetus to the development of perovskite solar cells. Almost all the materials applied inorganic photovoltaics are compatible in perovskite solar cells [21], but spiro-MeOTAD-based HTMs are still predominant. As the perovskite is electron conductive, the HTM thickness is important: too thick layer results in high series resistance; if the HTM layer is too thin, a short circuit between the back electrode and perovskite ensues. Generally, the thickness of HTM is controlled through a simple spin-coating process. Recently, HTM can even be dispensable benefited from the delicate design of perovskite materials with a longer exciton lifetime and a higher quantum yield for photoinduced charge separation [16].

5.1.4 Summary

Few years to its emergence notwithstanding, the perovskite solar cell strikes the world not only because of its arresting high efficiency but also its low cost and easy fabrication that outperform the conventional silicon solar cells. Branched from DSC and PSC, the working principal of the perovskite solar cell does not show many differences with its ancestors. So do the structures. In contrast to the increasing efficiencies recorded, the materials involved, including perovskites, HTMs and ETMs, are not diversified as expected. Most of the performance upgrades are attributed to the process control and interface engineering, which is more subtle than in DSCs and PSCs. Perovskite solar cells, the recruit in photoconversion devices, activated the new race to harness the sunlight, which paints a promising picture for the economical photovoltaics with PCE on par with the silicon solar cells.

5.2 Flexible Perovskite Solar Cell

As the popularity of wearable devices juices up the interest in flexible solar cells, there has been a lot of work relating the transformation towards flexibility including DSCs [22] and PSCs [23]. From a practical view, flexible solar cells should be compatible with solution-processed roll-to-roll printing fabrication method for large-scale production and application [24, 25]. Generally, a high-efficiency flexible DSC is fabricated in metal foil [26] or mesh [27, 28] as TiO_2 photoanode needs annealing in high temperature for recrystallization. For a flexible DSC built on a polyethylene terephthalate (PET) or polyethylene naphthalate (PEN) substrate, the efficiency is relatively low as the limitation of high-temperature annealing process and poor adhesion of TiO_2 and sealing problem looms large for long-term usage [22, 29]. The solid-state PSC is immune from the high-temperature treatment and sealing vulnerability, but its low efficiency and poor stability also inhibit its application [30].

Perovskite solar cells, endowing with solid state and high efficiency, have attracted increasing interest in transformation towards flexibility. But a series of tricky issues arise. Generally, perovskite solar cell contains a compact TiO_2 blocking layer and mesoscopic TiO_2 scaffold which requires annealing at high temperature. In this case, polymer substrates are ruled out [31]. And such rigorous processing conditions are not beneficial for large-scale production. Many efforts have been devoted to developing low-temperature process like using low-temperature TiO_2 [32], ZnO [33], graphene/TiO_2 composite [34], and Al_2O_3 [35]. For example, Mathews et al. applied ZnO compact layer and nanorod instead of TiO_2 to make flexible perovskite solar cells, which allowed low-temperature, solution-based processing (Fig. 5.4a) [36]. A PCE of 2.62 % was achieved on the flexible substrate. However, the open circuit is relatively low which suggested the

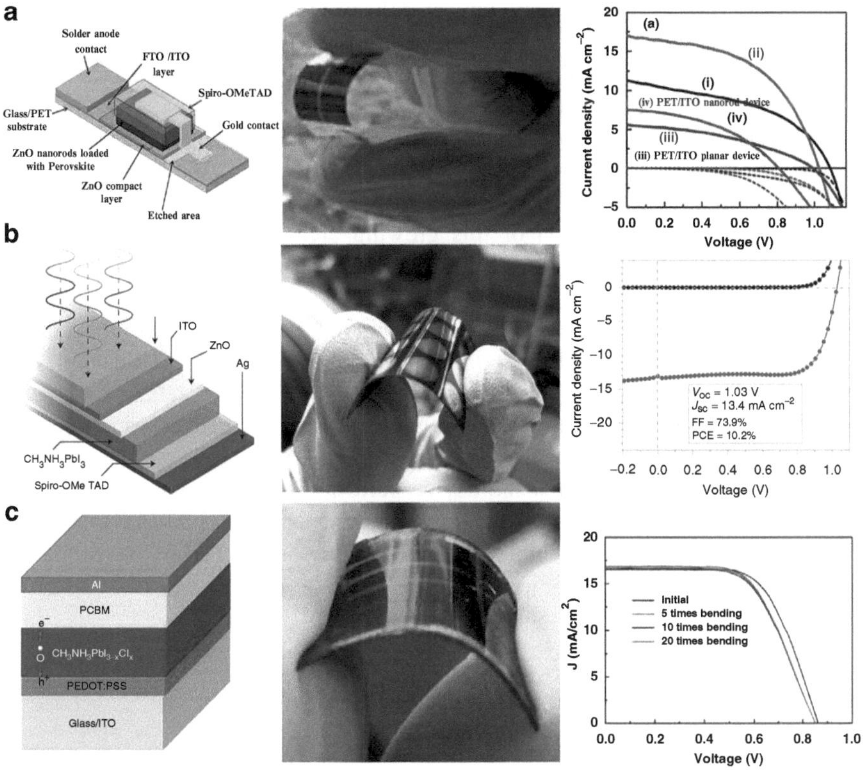

Fig. 5.4 Flexible perovskite solar cell in **a** mesoscopic structure (Reproduced from Ref. [36] by permission of The Royal Society of Chemistry). **b** Planar structure and **c** compatible with polymer solar cell structure and modification materials. **b** is reprinted by permission from Nature Publishing Group: Ref. [33], copyright 2014, and **c** is reprinted with the permission from Ref. [37]. Copyright 2014 American Chemical Society

severe charge recombination due to large surface area of nanorods and low perovskite coverage. Using low-temperature-operated ZnO compact layer, Kelly et al. successfully constructed a high-efficiency perovskite solar cell with conversion efficiencies of 15.7 % and 10.3 % for rigid and flexible substrate, respectively (Fig. 5.4b) [33].

Many strategies like materials selection, modification, and morphology control can learn from organic photovoltaics. In conventional PSCs, HTMs and ETMs are all processed at a low temperature and compatible with low-cost solution process. NiO, V_2O_5, poly(3,4-ethylenedioxythiophene)/poly(styrenesulphonate) (PEDOT: PSS), Spiro-OMeTAD, and other polymer materials can function as hole transporting layer and [6,6]-phenyl-C_{61}-butyric acid methyl (PCBM), TiO_2, and ZnO as electron transporting layer [31]. Snaith et al. demonstrated that the $CH_3NH_3PbI_{3-x}Cl_x$ perovskite absorber can be prepared from solution and sandwiched between the conventional charge collection layers and electrode configuration for organic photovoltaic diodes. A device in planar heterojunction structure consisting of fluorine-doped tin oxide (FTO) glass, PEDOT:PSS, perovskite absorber, PCBM, TiO_x, and aluminum can deliver a PCE of 10 %. When the layout was transferred to a flexible indium-doped tin oxide (ITO) plastic foil, a PCE of 6.5 % was recorded [31]. Similar work was also reported by Yang et al. with an efficiency of 9.2 % (Fig. 5.4c) [37].

For large-scale production of solar cell using roll-to-roll process, each layer requires flexibility and full printing operation possibility. While the cost and complex vacuum deposition needed for transparent electrode and metal electrode limit the practical application of this growing area, efforts on this aspect are undergoing with other alternative electrode materials like carbon black [16], carbon nanotubes [38], etc.

However, the perovskite is a crystalline material, hence cracks or other imperfection may occur upon bending it. Bolink et al. fabricated a flexible solar cell on a PET substrate and tested its bending endurance. A PCE of 7 % retained when the solar cell was bent for 50 times. No significant performance deterioration was observed indicating that the perovskite is compatible with roll-to-roll processing [39]. Further tests showed that the device performed well even after being bent repeatedly to decreasing radii of curvature. Importantly, Kelly et al. suggested that the flexibility of the device is limited by the flexibility of the ITO/PET substrate and not by the perovskite layer [33].

5.3 Fiber-Shaped Perovskite Solar Cell

With the popularity of wearable and portable devices, planar flexible energy devices will hardly keep up with surging demand. Electronic textiles are highly desired in many fields. Combining solid state and high efficiency, perovskite solar cells are promising in fiber-shaped photovoltaics. In this section, we present an effective method to transform perovskite solar cell into a flexible fiber shape [40].

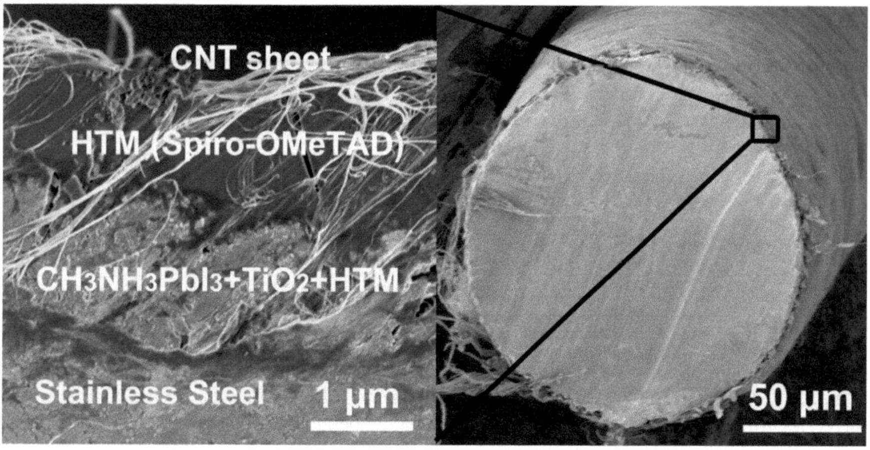

Fig. 5.5 Cross-sectional SEM image of a fiber-shaped perovskite solar cell (Reproduced from Ref. [40] by permission of John Wiley & Sons Ltd)

The fiber-shaped perovskite solar cell exhibited as coaxial configuration where functional layer was coated layer by layer. A stainless steel wire is chosen as the anode due to an appropriate work function and high electrical conductivity. In addition, the smooth surface has good affinity with a uniform compact TiO_2 blocking layer that is prepared from a titanium diisopropoxide bis(acetylacetonate) solution. And a mesoporous TiO_2 layer was subsequently coated to take in perovskite materials. Considering the difficulty of coating uniform thin perovskite film via dip-coating process, the mesoporous TiO_2, served as scaffold and ETM, was necessary. At last, HTM was coated on the perovskite layers followed by wrapping aligned CNT sheet as counter electrode (Fig. 5.5).

The transparent-aligned CNT sheet played a significant role in hunting high efficiency. In planar configuration, the devices are illuminated from the transparent conductive glass. The loss of illumination intensity is negligible. Differently, the coaxial fiber-shaped devices are illuminated from the counter electrode. The high transparent and flexible CNT sheet is an ideal material. A CNT sheet with optical transmittance of ~90 % ensures a high light absorption of photoactive materials. In addition, the difficulty of coating a uniform thin Ag or Au layer on curved surface highlighted the benefit of aligned CNT sheets.

The compact layer has significant influence on the final results. The annealing temperature affected the performance. A higher annealing temperature favored the formation of crystal n-type compact TiO_2 layer for higher efficiency while possibly produced more iron oxide produced from the oxidation of stainless steel to decrease the electron transport with lower efficiency (Fig. 5.6a). Furthermore, a thicker layer increases the series resistance that leads to a lower fill factor; the short-circuit current density is decreased with the increasing thickness of compact layer. In addition, thicker films became fragile and easily fracture into many stacked pieces. Therefore, it cannot effectively separate the steel wire and HTM, and a short circuit

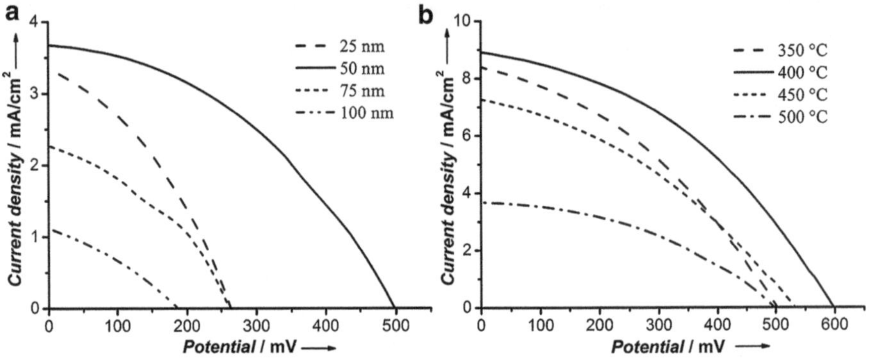

Fig. 5.6 J-V curves of fiber-shaped perovskite solar cells by varying the n-type compact TiO_2 layer. **a** Different thicknesses. **b** Different annealing temperatures (Reproduced from Ref. [40] by permission of John Wiley & Sons Ltd)

occurred. On the other hand, it is challenging to obtain very thin but continuous films due to the unavoidable existence of defects, which may also result in the short circuit. Here an optimized thickness is 50 nm (Fig. 5.6b). After optimization, the maximal PCE was obtained at the temperature of 400 °C. The PCE value of photovoltaic was 3.3 % (V_{OC}=0.664 V, J_{SC}=10.2 mA cm^{-2}, FF=0.487), which is insusceptible to incident angle and bending times (Fig. 5.7).

The fiber-shaped perovskite solar cells take the advantages of the sufficient contact and large interface of layer-by-layer configuration; hence, electrons and holes are extracted and diffused effectively (Fig. 5.5). A high PCE was obtained, in line with the excellent property of perovskite. In addition, coaxial structure was endowed with the stability of three-dimensional illumination and flexibility.

5.4 Stretchable Fiber-Shaped Perovskite Solar Cell

Considering the practical application, perovskite solar cells are further required to be integrated into elastic substrate to sustain stretch deformation. Elastic fiber-shaped photovoltaics, besides perovskite solar cell, succeed in fabrication but are still in the early stages of development.

The fabrication of the stretchable fiber-shaped perovskite solar cell is schematically shown in Fig. 5.8a. A Ti wire has been first shaped into a springlike structure, followed by growing a TiO_2 layer on the surface. Perovskite materials and HTMs including spiro-MeOTAD, lithium bis(trifluoromethane) sulfonamide (LiTFSI), and 4-tert-butylpyridine were dip coated on the modified springlike Ti wire, respectively. An elastic silicone rubber fiber wrapped with aligned CNT sheet was inserted into the springlike modified Ti wire, followed by wrapping with

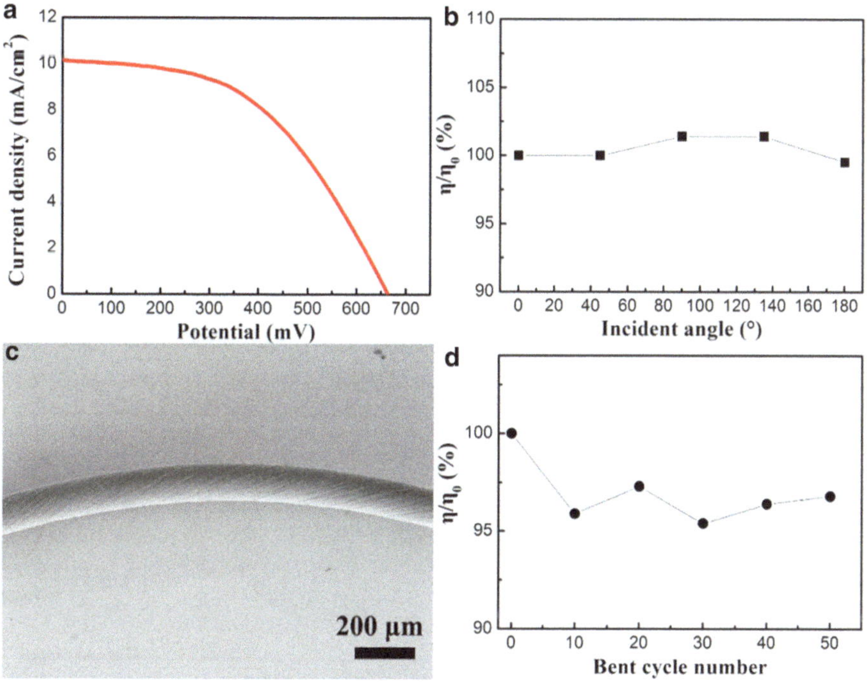

Fig. 5.7 **a** J-V curve of a coaxial fiber-shaped perovskite solar cell with energy conversion efficiency of 3.3 % (V_{OC}=0.664 V, J_{SC}=10.2 mA cm^{-2}, FF=0.487). **b** Dependence of energy conversion efficiency on the angle of incident light. Here η_0 and η correspond to the energy conversion efficiencies at 0 and the other angle, respectively. **c** SEM image of a bent fiber-shaped perovskite solar cell. **d** Dependence of energy conversion efficiency on bent cycle number. Here η_0 and η correspond to the energy conversion efficiencies before and after bending, respectively (Reproduced from Ref. [40] by permission of John Wiley & Sons Ltd)

another aligned CNT sheet to produce the stretchable fiber-shaped perovskite solar cell. Owing to the adequate bandgap, the electron and hole are able to separate and diffuse (Fig. 5.8b). In addition, the aligned CNT sheet stably attached on the HTM by the van der Waals interaction, facilitating the carrier collection (Fig. 5.8c) (unpublished work).

Various TiO_2 nanomaterials have been widely studied for planar perovskite solar cells, and TiO_2 nanotubes and nanoparticles are here compared for the fiber-shaped perovskite solar cell. Both TiO_2 nanotubes and nanoparticles offered high specific surface areas for absorbing photoactive materials, and $CH_3NH_3PbI_{3-x}Cl_x$ can be further infiltrated into the bottom of TiO_2 materials (Fig. 5.9). Importantly, perovskite materials were uniformly coated on and had fully covered the TiO_2 layer, which was key to a high photovoltaic performance. Compared with a planar surface, it is generally much more difficult to produce continuous and uniform films on curved surfaces. Here the high coverage of the perovskite material on the

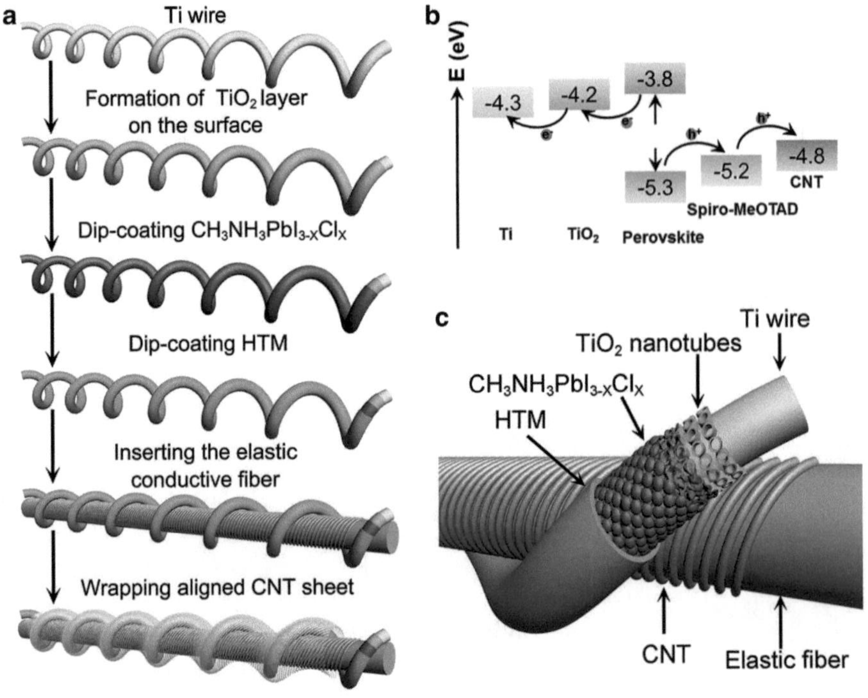

Fig. 5.8 Schematic illustration to the elastic PSC fiber. **a** Fabrication process. **b** Energy level diagram. **c** Structure

fiber-shaped surface was attributed to the presence of 1,8-diiodooctane that chelated with Pb^{2+} ions during the crystal growth, favoring the formation of high-quality thin perovskite films. Similarly, a uniform hole transport layer had been also achieved on the perovskite film (Fig. 5.9d, e). The CNTs remained highly aligned after the CNT sheet was wrapped on the elastic rubber fiber (Fig. 5.9g). A single CNT sheet that was covered on the twisted modified Ti wire and elastic conductive fiber maintained the continuous structure and stably bundled the two fiber electrodes together (Fig. 5.9f, h, i).

TiO_2 nanotube is a superior nanostructure for elastic fiber-shaped perovskite solar cells. The optimized length of TiO_2 nanotubes exhibited the PCE of 1.01 % (Fig. 5.10a), compared with the optimized PCE of 0.31 % for TiO_2nanoparticle-based device (Fig. 5.10b). This phenomenon may be explained by the fact that TiO_2 nanoparticles could not be uniformly deposited on the surface on the spiral fiber, in contrast to TiO_2 nanotubes grown with homogeneous height. In addition, fast electron transport property of TiO_2 nanotube further attributed to the enhancement of PCE. Furthermore, TiO_2nanotube-based devices also exhibited better bending and stretching performance. Figure 5.10c, d further compare the dependence of energy conversion efficiency on bent cycle number and strain in the stretchable

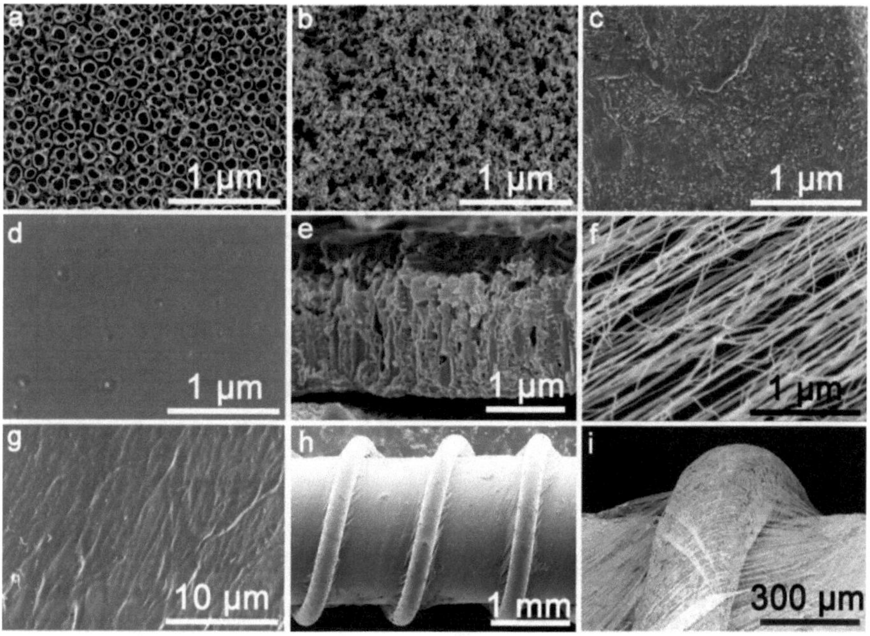

Fig. 5.9 SEM images of elastic PSC fibers. **a–d** *Top view* of aligned TiO_2 nanotubes, mesoporous TiO_2 nanoparticles, $CH_3NH_3PbI_{3-x}Cl_x$, and hole transport layer, respectively. **e** Cross-sectional image of the TiO_2 nanotube array coated with $CH_3NH_3PbI_{3-x}Cl_x$ and hole transport layer. **f, g** CNT sheet wrapped on modified Ti wire and elastic fiber, respectively. **h** An elastic PSC fiber with pitch distance of 1.25 mm. **i** High magnification of **h**

fiber-shaped perovskite solar cells based on TiO_2 nanotubes and nanoparticles. The energy conversion efficiencies had been maintained above 80 % and 25 % for TiO_2 nanotubes and nanoparticles under bending for 300 cycles, respectively; the energy conversion efficiencies were maintained above 90 % and 50 % for TiO_2 nanotubes and nanoparticles at a strain of 30 %, respectively. The elasticity had been further investigated by stretching the stretchable fiber-shaped perovskite solar cells at a strain of 30 % for 250 cycles, and the energy conversion efficiency was maintained by 90 % (Fig. 5.11a).

To satisfy various external devices, the stretchable fiber-shaped perovskite solar cells can be easily woven into textiles, and the perovskite solar cells can be connected in series or parallel to tune the output voltage or current, respectively. For instance, the output voltages had been increased from 0.63 to 1.88 V when three perovskite solar cells were organized in series, while the output currents could be enhanced from 156 to 412 μA when they were connected in parallel (Fig. 5.11c). As expected, as individual perovskite solar cells were flexible and elastic (Fig. 5.11b), the resulting perovskite solar cell textile was also flexible and elastic (Fig. 5.11d), which favored the application in flexible electronic devices.

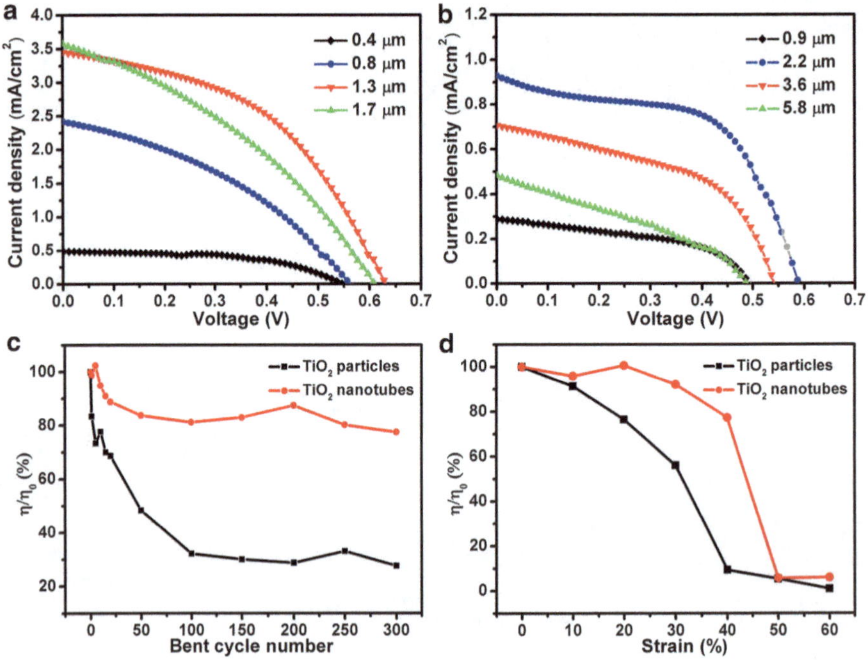

Fig. 5.10 J-V curves of elastic PSC fibers by varying the thickness of TiO_2 nanotubes **a** and mesoporous TiO_2 crystal **b.** Dependence of energy conversion efficiency on bent cycle number **c** and stretching strain **d.** Here η_0 and η correspond to the energy conversion efficiencies before and after bending or stretching, respectively

5.5 Perspective

The emergence of perovskite solar cells will absolutely bring about a revolution of photovoltaics in the next few years. The diversity of materials, including ETM, perovskite, and HTM, is the tendency of development currently and in a few years, aiming at higher PCEs. In the near future, over 20 % or even higher PCEs are highly feasible, and it is also available to be integrated with silicon or copper indium gallium selenide solar cells for achieving ~30 %. However, the PCEs are not the crucial factor for commercialization. The toxicity of Pb is one of the major impediments for application. In addition, the researches of stability are in the preliminary stage and gradually attracted widespread attentions. To this end, for further investigation, scientific researches should focus on more practical characterization by exquisite device design and targeted materials application, avoiding blindly hunting for high PECs.

The functionalization and integration of perovskite solar cell is another tendency, such as flexible and transparent devices. Furthermore, the fiber-shaped perovskite is another direction for functionalization. The all-solid-state micro-devices exhibited the energy conversion efficiency of ~3 %. It is not high enough,

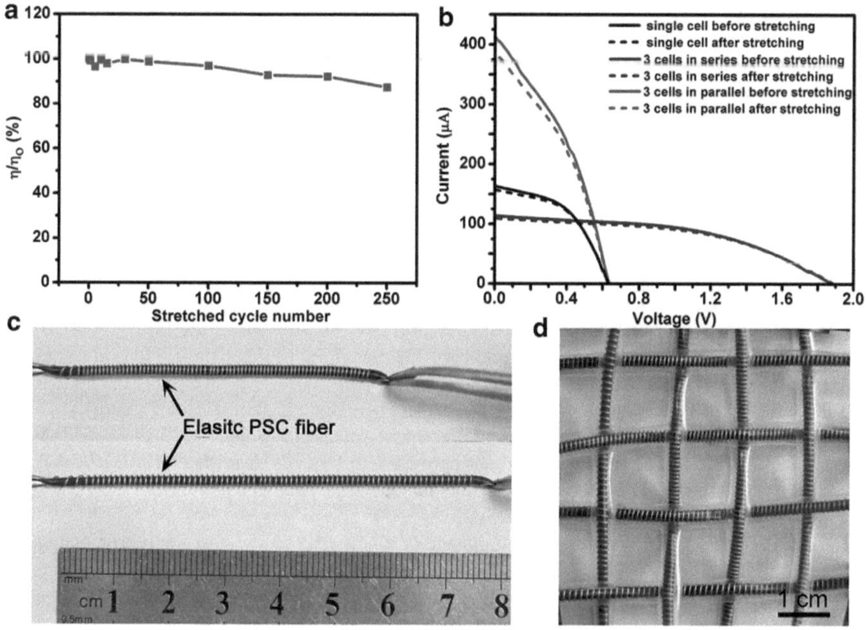

Fig. 5.11 **a** Dependence of energy conversion efficiency on stretched cycle number. **b** J-V curves of the resulting powering textile where three PSC fibers were connected in series or parallel before and after stretching. **c** Photographs of an elastic PSC fiber before and after stretching, respectively. **d** Photograph of an elastic powering PSC textile

but satisfactory, considering the first time to realize. The fiber-shaped perovskite solar cells with the unique shape and the characteristic of all-solid-state have unprecedented potential to produce energy fibers, even energy textiles. The development of novel coating process to fabricate uniform layers, similar as spin-coating process counterpart in planar solar cell, is key to boost the PECs fiber-shaped device. On the other hand, the fiber-shaped perovskite solar cells with endurance to deformation were achieved recently. It is an initial attempt to cater to the practical demand.

References

1. Mitzi D, Wang S, Feild C, Chess C, Guloy A (1995) Conducting layered organic-inorganic halides containing<110>-oriented perovskite sheets. Science 267(5203):1473–1476
2. Kojima A, Teshima K, Miyasaka T, Shirai Y (2006) Novel photoelectrochemical cell with mesoscopic electrodes sensitized by lead-halide compounds (2). In: Meeting abstracts, The Electrochemical Society, pp 397–397
3. Kojima A, Teshima K, Shirai Y, Miyasaka T (2009) Organometal halide perovskites as visible-light sensitizers for photovoltaic cells. J Am Chem Soc 131(17):6050–6051

4. Im J-H, Lee C-R, Lee J-W, Park S-W, Park N-G (2011) 6.5% efficient perovskite quantum-dot-sensitized solar cell. Nanoscale 3(10):4088–4093
5. Kojima A, Teshima K, Shirai Y, Miyasaka T (2008) Novel photoelectrochemical cell with mesoscopic electrodes sensitized by lead-halide compounds (11). In: Meeting abstracts, The Electrochemical Society, pp 27–27
6. Bach U, Lupo D, Comte P, Moser JE, Weissortel F, Salbeck J, Spreitzer H, Gratzel M (1998) Solid-state dye-sensitized mesoporous TiO_2 solar cells with high photon-to-electron conversion efficiencies. Nature 395(6702):583–585
7. Kim H-S, Lee C-R, Im J-H, Lee K-B, Moehl T, Marchioro A, Moon S-J, Humphry-Baker R, Yum J-H, Moser JE (2012) Lead iodide perovskite sensitized all-solid-state submicron thin film mesoscopic solar cell with efficiency exceeding 9%. Sci Rep 2:591
8. Docampo P, Guldin S, Leijtens T, Noel NK, Steiner U, Snaith HJ (2014) Lessons learned: from dye-sensitized solar cells to all-solid-state hybrid devices. Adv Mater 26(24):4013–4030
9. Lee MM, Teuscher J, Miyasaka T, Murakami TN, Snaith HJ (2012) Efficient hybrid solar cells based on meso-superstructured organometal halide perovskites. Science 338(6107):643–647
10. Zhou H, Chen Q, Li G, Luo S, Song T-b, Duan H-S, Hong Z, You J, Liu Y, Yang Y (2014) Interface engineering of highly efficient perovskite solar cells. Science 345(6196):542–546
11. Grätzel M (2014) The light and shade of perovskite solar cells. Nat Mater 13(9):838–842
12. Sum TC, Mathews N (2014) Advancements in perovskite solar cells: photophysics behind the photovoltaics. Energy Environ Sci 7(8):2518–2534
13. Snaith HJ (2013) Perovskites: the emergence of a new era for low-cost, high-efficiency solar cells. J Phys Chem Lett 4(21):3623–3630
14. Snaith HJ (2010) Estimating the maximum attainable efficiency in dye-sensitized solar cells. Adv Funct Mater 20(1):13–19
15. Liu M, Johnston MB, Snaith HJ (2013) Efficient planar heterojunction perovskite solar cells by vapour deposition. Nature 501(7467):395–398
16. Mei A, Li X, Liu L, Ku Z, Liu T, Rong Y, Xu M, Hu M, Chen J, Yang Y (2014) A hole-conductor–free, fully printable mesoscopic perovskite solar cell with high stability. Science 345(6194):295–298
17. Stranks SD, Eperon GE, Grancini G, Menelaou C, Alcocer MJ, Leijtens T, Herz LM, Petrozza A, Snaith HJ (2013) Electron-hole diffusion lengths exceeding 1 micrometer in an organometal trihalide perovskite absorber. Science 342(6156):341–344
18. Eperon GE, Stranks SD, Menelaou C, Johnston MB, Herz LM, Snaith HJ (2014) Formamidinium lead trihalide: a broadly tunable perovskite for efficient planar heterojunction solar cells. Energy Environ Sci 7(3):982–988
19. Hao F, Stoumpos CC, Cao DH, Chang RP, Kanatzidis MG (2014) Lead-free solid-state organic-inorganic halide perovskite solar cells. Nat Photonics 8(6):489–494
20. Noel NK, Stranks SD, Abate A, Wehrenfennig C, Guarnera S, Haghighirad A, Sadhanala A, Eperon GE, Pathak SK, Johnston MB (2014) Lead-free organic-inorganic tin halide perovskites for photovoltaic applications. Energy Environ Sci 7(9):3061–3068
21. Heo JH, Im SH, Noh JH, Mandal TN, Lim C-S, Chang JA, Lee YH, H-j K, Sarkar A, Nazeeruddin MK (2013) Efficient inorganic-organic hybrid heterojunction solar cells containing perovskite compound and polymeric hole conductors. Nat Photonics 7(6):486–491
22. Li L-L, Tsai C-Y, Wu H-P, Chen C-C, Diau EW-G (2010) Fabrication of long TiO_2 nanotube arrays in a short time using a hybrid anodic method for highly efficient dye-sensitized solar cells. J Mater Chem 20(14):2753–2758
23. Brabec CJ, Sariciftci NS, Hummelen JC (2001) Plastic solar cells. Adv Funct Mater 11(1):15–26
24. Krebs FC (2009) Polymer solar cell modules prepared using roll-to-roll methods: knife-over-edge coating, slot-die coating and screen printing. Sol Energy Mater Sol Cells 93(4):465–475
25. Youn H, Lee T, Guo LJ (2014) Multi-film roll transferring (MRT) process using highly conductive and solution-processed silver solution for fully solution-processed polymer solar cells. Energy Environ Sci 7(8):2764–2770

26. Liao JY, Lei BX, Chen HY, Kuang DB, Su CY (2012) Oriented hierarchical single crystalline anatase TiO_2 nanowire arrays on Ti-foil substrate for efficient flexible dye-sensitized solar cells. Energy Environ Sci 5(2):5750–5757
27. Pan S, Yang Z, Chen P, Deng J, Li H, Peng H (2014) Wearable solar cells by stacking textile electrodes. Angew Chem Int Ed 53(24):6110–6114
28. Fan X, Wang FZ, Chu ZZ, Chen L, Zhang C, Zou DC (2007) Conductive mesh based flexible dye-sensitized solar cells. Appl Phys Lett 90(7):073501
29. Qiu L, Wu Q, Yang Z, Sun X, Zhang Y, Peng H (2014) Freestanding aligned carbon nanotube array grown on a large-area single-layered graphene sheet for efficient dye-sensitized solar cell. Small. doi:10.1002/smll.201400703
30. Jorgensen M, Norrman K, Gevorgyan SA, Tromholt T, Andreasen B, Krebs FC (2012) Stability of polymer solar cells. Adv Mater 24(5):580–612
31. Docampo P, Ball JM, Darwich M, Eperon GE, Snaith HJ (2013) Efficient organometal trihalide perovskite planar-heterojunction solar cells on flexible polymer substrates. Nat Commun 4:2761–2766
32. Wojciechowski K, Saliba M, Leijtens T, Abate A, Snaith HJ (2014) Sub-150 °C processed meso-superstructured perovskite solar cells with enhanced efficiency. Energy Environ Sci 7(3):1142–1147
33. Liu DY, Kelly TL (2014) Perovskite solar cells with a planar heterojunction structure prepared using room-temperature solution processing techniques. Nat Photonics 8(2):133–138
34. Wang JT, Ball JM, Barea EM, Abate A, Alexander-Webber JA, Huang J, Saliba M, Mora-Sero I, Bisquert J, Snaith HJ, Nicholas RJ (2014) Low-temperature processed electron collection layers of graphene/TiO_2 nanocomposites in thin film perovskite solar cells. Nano Lett 14(2):724–730
35. Ball JM, Lee MM, Hey A, Snaith HJ (2013) Low-temperature processed meso-superstructured to thin-film perovskite solar cells. Energy Environ Sci 6(6):1739–1743
36. Kumar MH, Yantara N, Dharani S, Graetzel M, Mhaisalkar S, Boix PP, Mathews N (2013) Flexible, low-temperature, solution processed ZnO-based perovskite solid state solar cells. Chem Commun 49(94):11089–11091
37. You J, Hong Z, Yang YM, Chen Q, Cai M, Song TB, Chen CC, Lu S, Liu Y, Zhou H, Yang Y (2014) Low-temperature solution-processed perovskite solar cells with high efficiency and flexibility. ACS Nano 8(2):1674–1680
38. Li Z, Kulkarni SA, Boix PP, Shi E, Cao A, Fu K, Batabyal SK, Zhang J, Xiong Q, Wong LH, Mathews N, Mhaisalkar SG (2014) Laminated carbon nanotube networks for metal electrode-free efficient perovskite solar cells. ACS Nano 8(7):6797–6804
39. Roldán-Carmona C, Malinkiewicz O, Soriano A, Mínguez Espallargas G, Garcia A, Reinecke P, Kroyer T, Dar MI, Nazeeruddin MK, Bolink HJ (2014) Flexible high efficiency perovskite solar cells. Energy Environ Sci 7(3):994–997
40. Qiu L, Deng J, Lu X, Yang Z, Peng H (2014) Integrating perovskite solar cells into a flexible fiber. Angew Chem Int Ed 53(39):10425–10428

Chapter 6
Fiber-Shaped Supercapacitor

Abstract This chapter summarizes the development and applications of the fiber-shaped supercapacitors. The working mechanism and progress of supercapacitors are demonstrated in the beginning. Two architectures of supercapacitors, twisting and coaxial structures, are then discussed based on various fiber electrodes, active materials, and fabrication processes. Their performances including flexibility, stability, and specific capacitance are carefully illustrated. In addition, stretchable fiber-shaped supercapacitors reported in recent years are highlighted.

6.1 Overview of Supercapacitor

The rapid development of hybrid electric vehicles and portable electronic devices has aroused the interest in low-cost and environmentally friendly energy conversion and storage devices. Among them, lithium ion batteries and supercapacitors are on the cutting-edge list of high-performance energy storage devices as illustrated in the Ragone plot (Fig. 6.1) [1–5]. Ragone plot is a two-logarithmic coordinate chart used to map the performance of energy storage devices. Conceptually, the horizontal axis describes how much energy is available as indexed by specific energy (Wh kg^{-1}), while the vertical axis shows how quickly that energy can be delivered, represented by specific power (W kg^{-1}). Specific energy and specific power are two most important indexes to gauge an energy storage device. In some cases, volumetrically specified energy and power are also referenced, which are denoted as energy density (Wh L^{-1}) and power density (W L^{-1}). However, the difference between specific energy (or specific power) and energy density (or power density) is not that well defined, and it is often the case that the two indexed are blurred, and commonly, energy density (or power density) widely refers to the specific energy (or specific power). In this chapter, we have unified our expression by using energy density and power density to indicate gravimetrically specified energy and power. Compared with the commercial lithium ion batteries which possess high energy density as high as 180 Wh kg^{-1}, supercapacitors are highlighted for their much higher power density (10^5 W kg^{-1}), outstanding reversibility, and ultralong cycle life ($>10^5$ cycles).

H. Peng, *Fiber-Shaped Energy Harvesting and Storage Devices*, Nanostructure Science and Technology, DOI 10.1007/978-3-662-45744-3_6

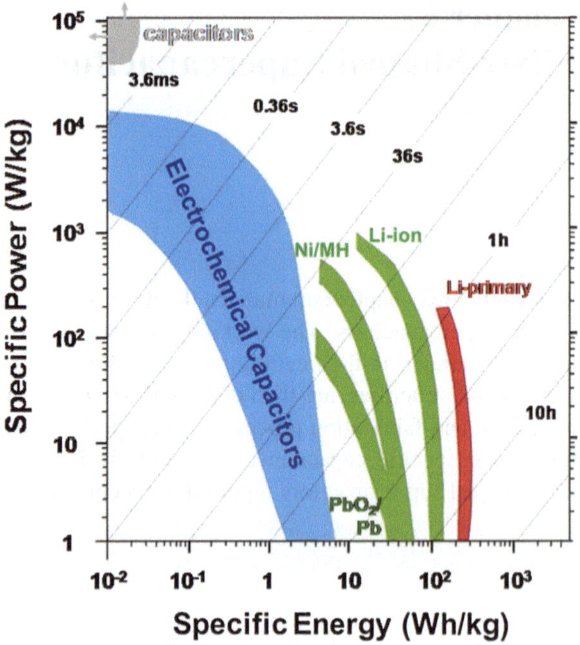

Fig. 6.1 Ragone plot of energy storage devices (Reproduced from Ref. [1] by permission of John Wiley & Sons Ltd.)

6.1.1 Energy Storage Mechanism

Supercapacitors, also scientifically named as electrochemical capacitors, are attracting much attention in a wide range of applications pertaining to electrical or hybrid vehicles, renewable energy generation devices, backup energy sources, and portable devices. According to the energy storage mechanism, supercapacitors can be classified into two categories: electrochemical double-layer capacitors where the capacitance stems from ion adsorptions at electrode/electrolyte interfaces and pseudocapacitors where the capacitance mainly comes from reversible redox reactions between electrolytes and electrochemical active species (Fig. 6.2) [2, 4, 6, 7]. In the electrochemical double-layer capacitor, an electrostatic field is built at the interface between the electrode and electrolyte, arising from charges accumulated at the near-surface region of the electrode and that counterbalanced by anions or cations in the electrolyte. Energy is stored in the electrostatic field. In pseudocapacitors, there is a charge transfer across the interface during charge and discharge process leading to a Faradic process between the electrode and electrolyte.

Energy is stored in the form of chemical reaction. The prefix "pseudo-" is applied to indicate the different origination of capacitance from the classical double-layer capacitors. On this account, charges migrate more rapidly in double-layer capacitors than in pseudocapacitors because no interface transfer is involved

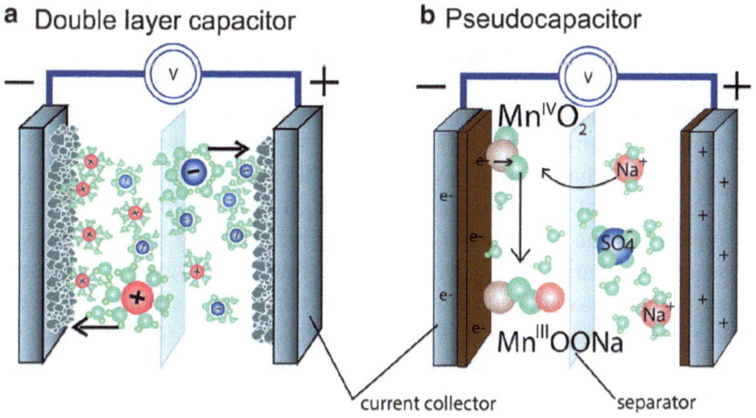

Fig. 6.2 Schematic illustration of two supercapacitors with different energy storage mechanisms. **a** Electrochemical double-layer capacitor. **b** Pseudocapacitor (Reproduced from Ref. [7] by permission of The Royal Society of Chemistry)

at the former process, indicating a larger power density and a better reversibility. Yet, the pseudocapacitors deliver a higher energy derived from the Gibbs free energy (ΔG) of chemical reactions. In many cases, these two mechanisms coexist in capacitors. The Faradic reactivity of surface functional moieties on electrodes contributes pseudocapacitance in double-layer capacitors, while in pseudocapacitors, the electrical double layers are ubiquitous at the interface, which brings in capacitance in the electrostatic field. The two electrodes in many capacitors are identical giving the name of symmetrical capacitor. Contrarily, there is an asymmetrical supercapacitor (also named hybrid supercapacitor) where electric double-layer capacitance and pseudocapacitance are integrated in different electrodes to bridge the gap between the massive energy storage and rapid energy delivery.

The electrochemical performances of supercapacitors are described by specific capacitance, energy density, power density, rate performance, and long cycle stability, which are commonly characterized by cyclic voltammetry and galvanostatic charge–discharge tests. The specific capacitance, the most important character of a supercapacitor indicating the capability of storing charges, is calculated from charge–discharge curve according to Eq. 6.1:

$$C_{\rm sp} = \frac{2i}{m\Delta V/\Delta t} \tag{6.1}$$

where m is the mass, which is usually specified to active materials, one electrode or entire device. The $C_{\rm sp}$ varies with different objects specified. In our view, the capacitance specified to entire device is more valuable in practical since it straightly exhibits the charge storage capability of a device. Academically, researchers also prefer other calculation methods to gauge electrochemical behaviors of different electrodes from the material design and structure engineering. i is the charge and discharge current and usually constant in the tests. ΔV is the voltage window applied during tests. The upper limit of voltage is determined by components in the electrolyte. Generally, a water-contained electrolyte has a voltage ceiling of 1.2 V (and usually charged to 1.0 V) which is the decomposition voltage of water. Organic electrolyte allows a higher charge voltage. In principle, the capacitance is impervious to voltage window but dictated by the scanning rate, $\Delta V/\Delta t$, which is the slope of the charge–discharge curve. As reflected in cyclic voltammogram (CV), assuming a constant capacitance, the CV curve always exhibits a rectangular shape. In fiber-shaped devices, length-specific capacitance (C_L) and area-specific capacitance (C_A) are also introduced for performance assessment, which are expressed as

$$C_L = \frac{2i}{L\Delta V/\Delta t} \tag{6.2}$$

$$C_A = \frac{2i}{A\Delta V/\Delta t} \tag{6.3}$$

where the L and A refer to the length and surface area of the electrode, respectively. Accordingly, C_L and C_A are superficial descriptions to the charge storage capability within a certain length and area. Many factors such as electrode diameter and microstructures relating to the capacitance are obscured. Hence, in fundamental research, C_L and C_A are recommended as auxiliary indexes for self-comparison. C_{sp} is still indispensable in fiber-shaped supercapacitors.

Energy density (E) and power density (P) of a capacitor are defined as

$$E = \frac{CV^2}{2} \tag{6.4}$$

$$P = \frac{V^2}{4R_s} \tag{6.5}$$

where R_S (Ω) is the equivalent internal resistance of the capacitor, corresponding to the series resistances of the electrolyte and electrode and contact resistance between the collector and electrode. Compared with lithium ion batteries, supercapacitors can accomplish charge and discharge processes within seconds, making them superior in energy delivery (~15 kW kg^{-1}). Though suffering a lower energy density, supercapacitors are expected to play a critical role in complementing or replacing batteries in the energy storage field, not least as backup supplies against power disruption and providing transient large power when needed. The energy density of a supercapacitor can still be elevated by enlarging the surface area, improving conductivity, and increasing the voltage (Fig. 6.3).

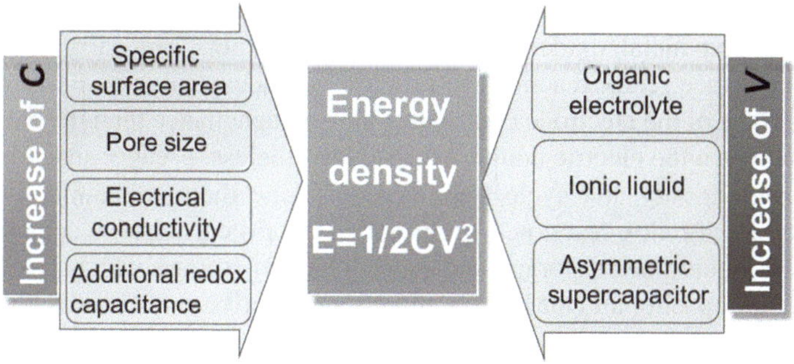

Fig. 6.3 Approaches towards high energy density of a supercapacitor (Reproduced from Ref. [1] by permission of John Wiley & Sons Ltd.)

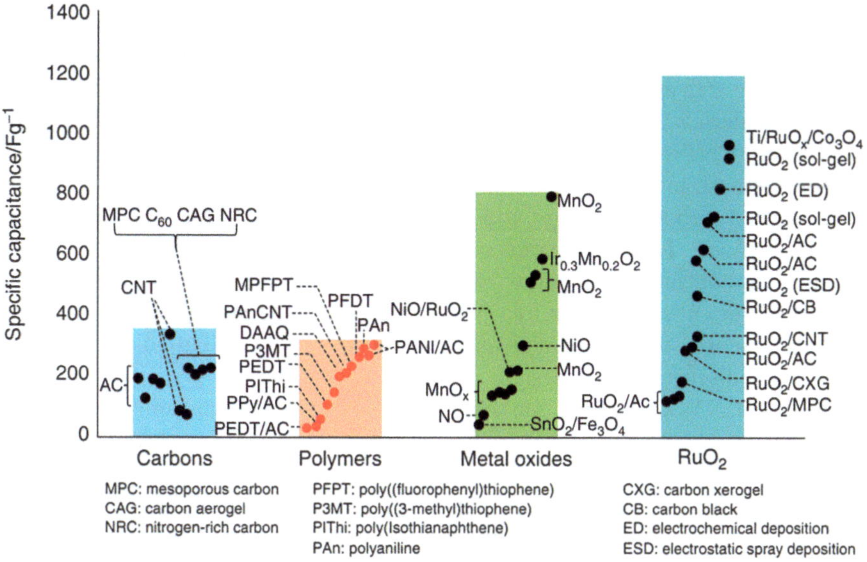

Fig. 6.4 Comparison of capacitive performances of different materials in supercapacitors (Reprinted with permission from Ref. [13]. Copyright 2008, The Electrochemical Society)

6.1.2 *Electrode Material*

Supercapacitors typically comprise three parts: two electrodes sandwiching the electrolyte and separator. Separators are not necessary since they are sometimes acted by gel electrolyte. It is undoubtedly that electrode materials play an essential role to the device performance since they reign charge storage and transportation. Reviewing the development of supercapacitors, the upgrade in performance is always accompanied by the revolution of electrode materials. After years of investigation, electrode materials are diversified. They can be categorized into carbon materials, transition metal oxides, and conducting polymers, as displayed in Fig. 6.4 [8–13].

Carbon materials can provide high surface area for electrostatic charge storage, while transition metal oxides and conducting polymers can carry out fast and reversible surface redox reactions to provide pseudocapacitance. Faradic process at the interface of the electrode contributes higher capacitance than the electrostatic charges between the electric double layers. Nevertheless, pseudocapacitors exhibit lower power density and cycle stability due to the thermodynamic and kinetic irreversibility of redox reaction. To this end, a hybrid capacitor, a combination of the electric double-layer storage and redox reaction storage which benefits from merits of the capacitor and the battery, directs an effective way towards high-energy and high-power energy storage devices.

6.1.2.1 Carbon Materials

In the early research of supercapacitors, carbon with large surface area was coated on a metallic current collector as the electrode. To achieve a high capacitance, a high surface area and remarkable conductivity are always desired for an electrode. Carbon materials such as activated carbon, carbon aerogels, graphite, carbon nanotubes (CNTs), and carbon nanofibers are satisfying candidates because of reasonable cost, low toxicity, and tailorable structure.

Activated carbon is the most widely used material in a commercial capacitor in liquid medium. They are produced from biomaterials or polymer precursors through carbonization and pore forming. However, the embedded surface in activated carbon impedes ion transport in the electrolyte at high voltage (>3 V), and microscopic defects of activated carbon are easily collapsed during rapid charge and discharge processes.

Instead, CNTs are coming into the sight. CNTs where carbon atoms in CNTs are sp^2 hybridized have been popularized as electrode materials for supercapacitors due to their high aspect ratio, large specific surface area, and excellent mechanical and electrical properties. CNTs can be prepared by chemical vapor deposition (CVD) from hydrocarbons (such as methane, acetylene, ethylene, etc.) as gaseous precursor. Compared to other carbon materials such as activated carbon, carbide-derived carbon, and zeolite-templated carbons, CNTs have higher conductivity and open-pore structure which favor rapid ion migration into the mesoporous structure. The double-layer capacitance of CNTs can reach 200 F g^{-1} which relies on structure, morphology, and purity of the CNT.

As early as 1997, Niu et al. reported HNO_3-treated multiwalled CNTs with an average pore size of 8 nm and a specific surface area of 430 m^2 g^{-1}, which exhibited a power density of 8 kW kg^{-1} and a specific capacitance of 102 and 49 F g^{-1} at 1 and 100 Hz in 38 wt% H_2SO_4 solution, respectively [14]. Frackowiaket et al. described the electrochemical behavior of multiwalled CNTs in KOH solution, which exhibited a maximum capacitance of 80 F g^{-1} [15]. An et al. found that single-walled CNTs delivered a higher specific capacitance up to 180 F g^{-1} in KOH aqueous solution with power density of 20 kW kg^{-1} and energy density of 7 Wh kg^{-1} [16]. Moreover, apart from large capacitance, the CNTs play a brilliant role in flexible and wearable devices, which will be presented in this chapter.

6.1.2.2 Transition Metal Oxides

Transition metal oxides such as MnO_2, RuO_2, and $Ni(OH)_2$ have been intensively studied as electrode materials. In these supercapacitors, reversible redox reactions take place providing large pseudocapacitance. RuO_2 is a popular electrode material because of its decent conductivity and ultrahigh pseudocapacitance (more than 600 F g^{-1}) derived from three distinct oxidation states accessible within 1.2 V. However, the high price of RuO_2 is not favorable for large-scale production, and the small voltage window limits its improvement in organic electrolytes [17–19]. Less expensive surrogates like MnO_2 are then investigated. But challenges to MnO_2 cannot be neglected: the low electron conductivity, low ion diffusion constant, and the manganese dissolution in the electrolyte solution [20–22].

Encouraging progresses have been made in tailoring MnO_2 nanostructures to overcome its disadvantages. The conductive additives, such as CNTs or graphene, are mixed with MnO_2 to enhance electrochemical performances. Yu et al. prepared paper-like electrode by electrochemically depositing MnO_2 on a flexible CNT paper and then incorporating with graphene [23]. Graphene sheets affixed on the surface enhanced the conductivity and prevented MnO_2 divorcing from CNT matrix. This ternary electrode (graphene/MnO_2/CNT) showed a higher specific capacitance (486.6 F g^{-1}) and cycle stability. In addition, MnO_2 is competent to be a pseudocapacitive positive electrode in hybrid supercapacitors. For example, when paired with CNT/polyaniline electrode, the hybrid supercapacitor showed a high energy density of 24.8 Wh kg^{-1} and a high voltage of 1.6 V.

Inspired by the fact that high surface area promises high capacitance, a variety of nanostructures of metal oxide are developed covering dendritic cluster, nanocrystals, nanoribbons, nanoneedles, nanotubes, etc. For example, Hou et al. [24] recently designed a rational structure by winding the conductive polymer, poly (ethylenedioxythiophene) (PEDOT), and MnO_2 to improve the overall performance. Lee et al. [25] adopted a layer-by-layer deposition method to composite MnO_2 with CNTs. Lang et al. [26] developed a hybrid metal oxide nanoporous structure with a specific capacitance up to 1,145 F g^{-1}, close to the theoretical capacitance of MnO_2. Hu et al. deposited MnO_2 on the CNT fabric to attain a three-dimensional (3D) network structure [27]. As shown in Fig. 6.5, the 3D structure can undertake a high loading of MnO_2 (8.3 mg cm^{-2}) without sacrificing the mechanical strength. The strong interaction between MnO_2 and CNT makes the particles uniformly and intimately affixed on the matrix. The fabric supercapacitor exhibited a capacitance up to 410 F g^{-1} in Na_2SO_4 aqueous solution.

6.1.2.3 Conducting Polymers

As shown in Fig. 6.4, electrochemically reactive polymers for supercapacitors are polyaniline (PANI), polypyrrole (PPy), poly(3,4-ethylenedioxythiophene) (PEDOT), polythiophene (PTh), and their derivatives [11, 21, 28]. Theoretically,

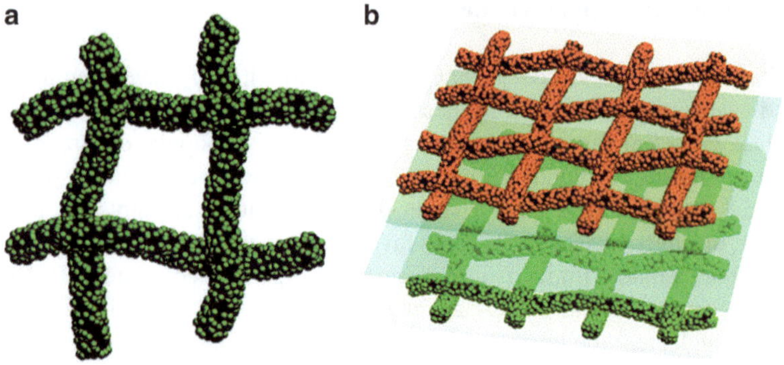

Fig. 6.5 **a** Schematic illustration to MnO_2-deposited CNT fabric. **b** A symmetric pseudocapacitor based on CNT/MnO_2 fabric. The positive electrode (*bottom*) is MnO_2, while the negative electrode (*top*) is reduced MnO_2 (Reprinted with the permission from Ref. [27]. Copyright 2011 American Chemical Society)

conducting polymers can deliver high pseudocapacitance which is calculated according to their redox reactions. For example, PANI has a theoretical capacitance of 1,284 F g^{-1}; PPy, 480 F g^{-1}; and PEDOT, 210 F g^{-1}. But the theoretical value is not accessible, even approachable, because the active materials in the internal electrode cannot be utilized completely. Besides, compared with pseudocapacitors based on metal oxides, conducting polymers exhibit less stability during long-term charge and discharge cycles.

To this end, CNTs are introduced to leap the hurdle of conducting polymers in electrical conductivity, mechanical strength, and thermal stability and provide a high specific surface area as matrix. Wei et al. used chemical oxidation to grow PANI nanowires on single-walled CNT cloth to fabricate a flexible supercapacitor [29]. It achieved a specific capacitance of 410 F g^{-1}, much larger than the individual single-walled CNTs (60 F g^{-1}) and PANI (290 F g^{-1}). The enhancement was attributed to the synergetic effect of single-walled CNTs and PANI that single-walled CNTs provide a conductive network, beneficial for consuming the active materials. More than 90 % of capacitance survived after 3,000 cycles of charge and discharge.

Apart from CNTs, other carbon materials, such as porous carbon, which possesses large surface area, are also reported to support conducting polymers. Mondal et al. deposited PANI on porous carbon and achieved a capacitance up to 1,600 F g^{-1} [30].

6.1.2.4 Fiber-Shaped Supercapacitor

The fiber-shaped supercapacitors which have one-dimensional (1D) configuration gain increasing popularity in applications pertaining to portable and wearable electronics. Recently, efforts have been devoted in fabricating fiber-shaped supercapacitors in twisting or coaxial structure. Technically, gel electrolytes are usually used to overcome the safety vulnerability of liquid electrolytes. In stark contrast to the conventional planar supercapacitor, fiber-shaped supercapacitors can be readily woven into textiles for flexible and wearable applications.

Fiber-shaped supercapacitors are constructed on suitable fiber electrodes. Various fiber electrodes have been explored, including CNT fibers, graphene fibers, CNT/conducting polymer composite fibers, and polymer/metal oxide composite fibers. Among these fiber electrodes, the CNT fiber displays many unique merits such as lightweight, excellent mechanical properties, and high thermal and electrical conductivity, making it outstanding in fiber-shaped supercapacitors.

As previously declared, solid-state electrolytes are favored as the safe electrolyte in twisting fiber-shaped supercapacitors [31]. A polymer gel electrolyte can serve as separator to prevent short circuit. Poly(vinyl alcohol) (PVA)/inorganic acid gel electrolyte is commonly used. The performance of the gel electrolytes in a fiber supercapacitor has been tested in a parallel comparison with a KNO_3 liquid electrolyte. PVA/H_3PO_4 electrolyte demonstrated a higher specific capacitance, which is more suitable than liquid electrolytes for fiber supercapacitors.

Besides designing novel structures, many efforts have been devoted to integrating functions to supercapacitors to extend their applications. For instance, they have been made flexible and stretchable to accommodate the deformation during use. The combined high flexibility and electrochemical property in fiber-shaped devices promise potential applications like electronic textiles.

6.2 Twisting Fiber-Shaped Supercapacitor

6.2.1 Overview

Twisting fiber electrodes into fiber-shaped devices is proved a reliable and effective strategy. Several representative works on twisting supercapacitors were summarized in Fig. 6.6 [2, 31–36]. The two fiber electrodes can be prepared independently. Active materials, such as transition metal oxides and conductive polymers, can be incorporated with fibers to introduce pseudocapacitance. The obtained supercapacitor exhibits a typical one-dimensional configuration whereupon can be knitted and weaved into a fabric. The electrolyte will infiltrate into the space between two electrodes avoiding the short circuit. Experimentally, the twisting electrodes are usually encased within a tube for protection.

For example, ZnO nanowires with large specific surface area benefit charge storage, which can be grown at low temperatures (less than 90 °C). MnO_2 was incorporated into the ZnO nanowire composite fiber to introduce pseudocapacitance. The fiber supercapacitors can be integrated with the fiber-shaped generators in wearable power system. Unfortunately, the power density and energy density of these supercapacitors were significantly lower compared to carbon-based supercapacitors [32].

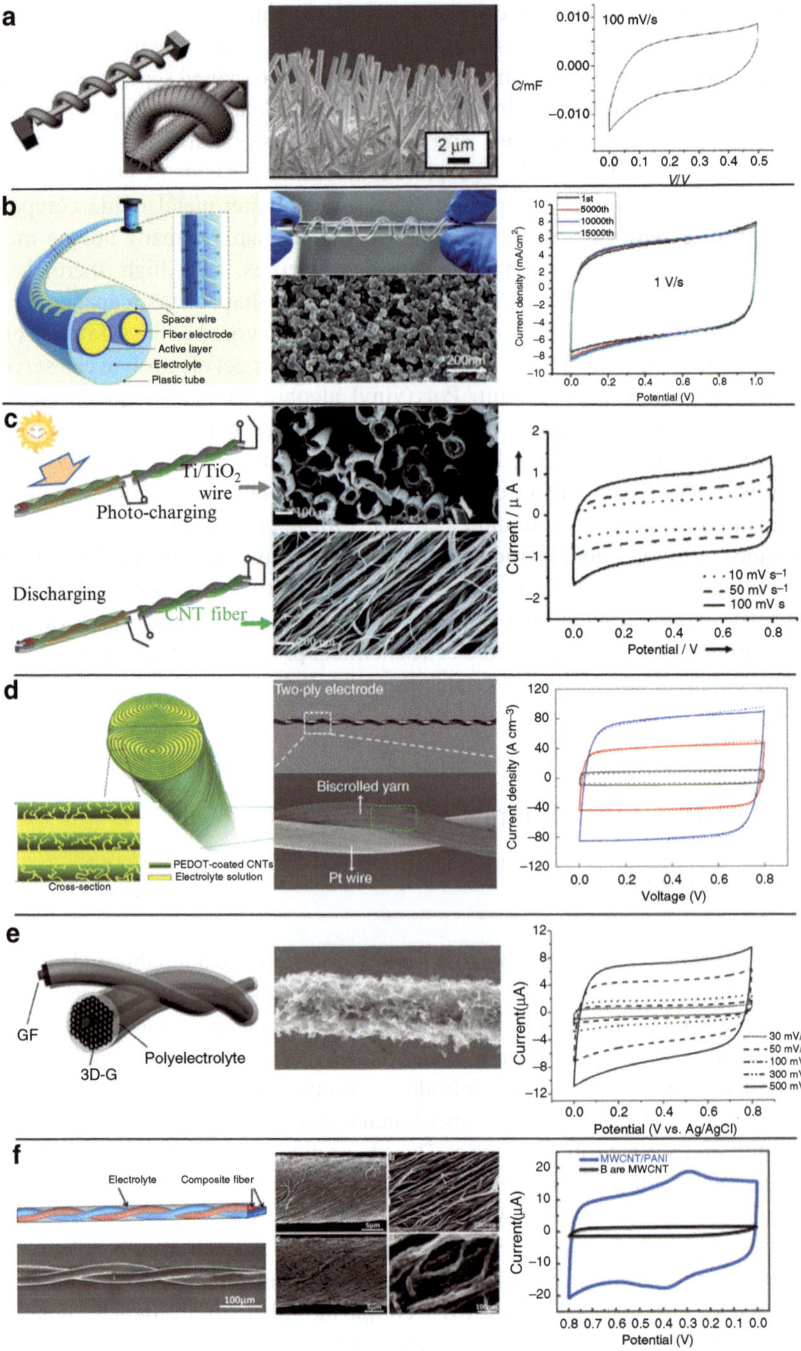

Fig. 6.6 Fiber-shaped supercapacitors in twisting structures. *Left row*, schematic illustrations; *middle row*, images of fiber electrodes; *right row*, typical cyclic voltammograms. **a**. ZnO nanowire coated on synthetic fibers (Reproduced from Ref. [32] by permission of John Wiley & Sons Ltd). **b**. Active carbon ink coated fiber (Reproduced from Ref. [2] by permission of John Wiley & Sons Ltd).

Zou et al. designed a fiber-shaped supercapacitor with pen ink coated on the surface of fiber electrodes [2]. The areal capacitance and energy density of the supercapacitor were 19.5 mF cm^{-2} and 2.70? × ?10^{-6} Wh cm^{-2}, respectively. The power density was 9.07 mW cm^{-2}. This fiber-shaped supercapacitor could be stably performed over 15,000 charge–discharge cycles. The fiber-shaped supercapacitor was sustainable for bending deformation without structure fracture and performance deterioration. Furthermore, this low-cost and commercially available materials and simple fabrication method paved a feasible road to large-scale production.

In the following sections, we will present a series of fiber-shaped supercapacitors based on a CNT fiber.

6.2.2 Carbon Nanotube Fiber/Ti Wire Hybrid Supercapacitor

Since the fiber electrode can be prepared independently in twisting structure, it is easy to fabricate hybrid supercapacitors, also known as asymmetric supercapacitors in some circumstances [33]. For example, in a novel energy wire, which integrated a fiber-shaped supercapacitor with a fiber-shaped organic solar cell, the supercapacitor shares a Ti wire, which was modified with aligned TiO_2 nanotubes, with the solar cell. TiO_2 nanotubes, as photoanode in solar cell, played a key role in harvesting sunlight after dye sensitization, while they improved the charge transport in the electrode and increased the energy storage area in the supercapacitor part. Aligned CNT fibers were twisted around the functional Ti wire to fabricate the multiple devices. The supercapacitor achieved a specific capacitance of 64.9 mF cm^{-1} or 1.65 mF cm^{-2} at a scan rate of 10 mV s^{-1}. However, the mismatching of capacitances of modified Ti wire and CNT fiber made the hybrid supercapacitor weakened in performance. This hybrid supercapacitor indicated that functional Ti wire with a diameter of 200 μm was far beyond the capacity of CNT fiber with a diameter of 10 μm. The power of supercapacitor part was reduced due to the mismatching. As a comparison, an integrated device from polymethylmethacrylate (PMMA) fibers showed a low energy-conversion efficiency of 0.02 % and specific capacitance of 0.4 mF cm^{-2}. The CNT fiber and functional Ti wire were further investigated as a supercapacitor by the successors. The wire devices had been woven as a textile, which could light up a red light-emitting diode (LED) lamp

Fig. 6.6 (continued) **c.** TiO_2-modified Ti wire wound by CNT fiber (Reproduced from Ref. [33] by permission of John Wiley & Sons Ltd). **d.** Biscrolled CNT/PEDOT:PSS composite fiber twisting with platinum wire (Reprinted by permission from Nature Publishing Group: Ref. [36], copyright 2013). **e.** All graphene core–sheath fibers (Reproduced from Ref. [35] by permission of John Wiley & Sons Ltd). **f.** CNT/PANI composite fiber (Reproduced from Ref. [34] by permission of The Royal Society of Chemistry)

either in harvesting or storage sate. A typical dye-sensitized solar cell was conducted by a redox ion couple between titanium dioxide nanotubes and CNTs. The ion transportation means an energy transfer between electrodes. This hybrid energy wire device complicated both energy harvesting and storage with a constant wire. The supercapacitor state also achieved a high specific capacitance of 64.9 mF cm^{-1} or 1.65 mF cm^{-2} at a scan rate of 10 mV s^{-1}. In contrast, the titanium dioxides were kept without the dye-sensitized procedure. The fiber-shaped device failed to be effectively charged and discharged without the dye. The reason was that the transportation of I_3^-/I^- ion couple could not carry on without the dye. Beyond the performance and safety of the energy storage state, the concept to produce one wire device with dual functions provided a novel strategy in the development of future electronics.

6.2.3 Carbon Nanotube Fiber/Conducting Polymer Composite Fiber

As previously mentioned, conducting polymers such as polyaniline (PANI), polypyrrole (PPy), and poly(3,4-ethylenedioxythiophene) (PEDOT) are widely used to deliver pseudocapacitance in supercapacitors [37]. Compositing with CNTs has been proved an effective way to enhance the mechanical strength and electrical conductivity. Hence, we developed a simple and general strategy to fabricate high-performance fiber-shaped supercapacitors by incorporating aligned CNT fiber with conducting polymers [34]. As a demonstration, PANI was deposited to the surface or near surface of CNT fiber through a solution-based electrochemical polymerization. The amount of PANI deposited was calculated through Faraday's law and can be varied by controlling the quantity of charge at constant voltage. The resulting fiber-shaped supercapacitor fabricated from twisting two composite fibers produced a specific capacitance of 274 F g^{-1} or 263 mF cm^{-1}. The high performance was attributed to the synergetic effect of the conducting polymer and CNTs. In the composite fiber, PANI intimately contacted with the CNTs, which provided a high specific surface area and efficient charge transfer pathways. The CNT/PANI composite supercapacitor showed excellent capacitance retention of 91 % after 800 charge and discharge cycles, indicating a remarkable cyclic stability. Moreover, the CNT framework in the fiber maintained the mechanical stability so that the composite fiber was insusceptible to the repeated deformation. No obvious structural damage was observed after bending for 100 times. Due to its high flexibility, the fiber-shaped supercapacitor can be easily woven into clothes by conventional textile technology.

Poly(3,4-ethylenedioxythiophene), which was well known to be highly Faradaically conducting and non-Faradaically stable in the oxidized state, was also deposited onto the CNT sheet by vapor phase polymerization. Polymer-infiltrated CNT yarns were scrolled into a fiber. The volumetric capacitance of the plied biscrolled yarn/Pt wire device is up to 179 F cm^{-3}. The complete supercapacitor could be wound or sewn for 10,000 cycles, while the performance retention rate was over 90 %. The exceptionally high cycle life was important for the applications. This solution-based processing method provides a route to the industrialized production of high-performance biscrolled yarn supercapacitors [36].

6.2.4 Carbon Nanotube Hybrid Fiber

As a popular functional carbon material, ordered mesoporous carbon (OMC), with tunable pore structures and sizes that promise a high surface area for high specific capacitance, has been incorporated with aligned CNTs to form a hybrid fiber [38, 39]. OMC was synthesized from SBA-15 as the template. The Brunauer–Emmett–Teller (BET) measurement provided the average size of 3.4 nm and specific surface area of 1,014 m^2 g^{-1}. The aligned CNT network with high electrical conductivity realized the rapid charge transportation, which also favored the charge collection from OMCs. The high specific surface area of OMC components is beneficial for charge accumulation. With the increasing content of OMC from 0 to 87 wt%, the specific capacitance and discharge area energy density were improved for 100 and 10 times, respectively. The specific capacitance was 39.7 mF cm^{-2} (or 1.91 mF cm^{-1}), and energy density reached $1.77?\times?10^{-6}$ Wh cm^{-2} (or $8.50?\times?10^{-8}$ Wh cm^{-1}). The output power density was up to $4.30?\times?10^{-2}$ mW cm^{-2} ($1.87?\times?10^{-3}$ mW cm^{-1}) at a current of 0.01 mA. The fiber-shaped supercapacitor retained 87 % of its specific capacitance after 500 cycles, indicating a decent long-life stability. Furthermore, the CNT/OMC hybrid fiber is flexible and lightweight, which is important for wearable electronics.

It has been exhibited that the internal porous structure in the electrode is beneficial for a high-performance supercapacitor [38]. Twisting multiwalled CNT yarns were endowed with high internal porosity, and MnO_2 was deposited as active material to introduce pseudocapacitance. The three-dimensional (3D) porosity inside the yarn enabled an enlarged space accessible for electrolyte. Moreover, the active materials were besieged with CNT frameworks, which, as a result, facilitate the utilization of active materials and make the electrode adaptable at high rate. Consequently, the supercapacitor delivered a specific capacitance of 25.4 F cm^{-3} at 10 mV s^{-1} and energy and power densities of 3.52 mWh cm^{-3} and 127 mW cm^{-3}, respectively.

Chen and Dai prepared a hierarchically structured carbon microfiber comprising interconnected aligned single-walled CNTs interposed with nitrogen-doped reduced graphene oxide sheets [40]. The hierarchically structured nanomaterial has a large specific surface area (396 m^2 g^{-1}) and high electrical conductivity (102 S cm^{-1}). The resultant fibers showed a specific capacity as high as 300 F cm^{-3} in PVA/H_3PO_4 electrolyte. More surprisingly, the supercapacitor delivered a volumetric energy density of ~6.3 mWh cm^{-3}, on par with a thin-film lithium battery while giving a power density more than two orders of magnitude higher than that of batteries, as well as a long cycle life. This fiber-shaped supercapacitor can be easily integrated into miniaturized flexible devices, such as powering an ultraviolet photodetector and an LED.

6.2.5 Graphene Fiber

Three-dimensional (3D) graphene possesses arresting features including highly exposed surface areas, high electrical conductivity, and good chemical stability, and therefore, they have been widely explored as electrode materials in supercapacitors [41]. A graphene-based core–sheath fiber, where graphene fiber was the core and 3D porous graphene framework was the sheath, was prepared though electrochemically electrolyzing [35]. The hybrid fiber integrated the high conductivity provided by the graphene fiber in the core with the large surface area donated by the 3D graphene framework, making itself competent as a flexible electrode in the fiber-shaped supercapacitor. Two hybrid fibers were intertwined to fabricate a symmetric supercapacitor immersed gel electrolyte (H_2SO_4/PVA). The assembled supercapacitor exhibited remarkable flexibility and managed to endure deformations when shaped and woven into textiles.

The core–sheath structure seems popular in designing the fiber-shaped supercapacitors. Gao and his coworkers designed a coaxial wet-spinning assembly strategy to prepare polyelectrolyte-wrapped graphene fibers [42]. The setup was delicately designed with a double-inlet spinneret. The graphene oxide (GO) was fed to the inner spinning dope, while the sodium carboxymethyl cellulose (CMC), an ionic conductor, was fed to the sheath. The wet-spinning approach produced a core–sheath fiber, denoted as GO@CMC fiber, with polyelectrolyte encasing the graphene oxide fiber, which prevented the twisting electrodes from short circuit and provided access for ion penetration. This strategy is compatible with various inner spinning dope, like CNT dispersion and CNT/RGO mixture. It can be generalized to produce different core–sheath fibers. Then, the GO@CMC fibers were chemically reduced by hydroiodic acid and the graphene oxide sheath was converted to reduced graphene oxide (RGO). The RGO@CMC had a conductivity of ~70 S cm^{-1}. The core–sheath hybrid fiber exhibited high electrochemical performance in the twisting fiber-shaped supercapacitor. For example, a supercapacitor with RGO + CNT@CMC fiber as electrodes had a capacitance of 177 mF cm^{-2} (158 F cm^{-3}) and energy density of 3.84 mWh cm^{-2} (3.5 mWh cm^{-3}). The polyelectrolyte sheath makes the supercapacitor exempt from the electrolyte leakage not least when the device is deformed. The core–sheath fiber was flexible and robust enough to sustain the long-term and repeated deformation, and the capacitance was slightly fluctuated during 1,000 times of bending.

In most cases, fiber-shaped supercapacitors are fabricated in a twisting structure where the two fiber electrodes are intertwined together. Recently, the development in fiber-shaped supercapacitor is burgeoning as fiber electrodes are diversified, derived from various carbon nanomaterials. The threshold of fabricating supercapacitor is much lower compared with its analogues, solar cells, and batteries that require complicated treatment and delicate skills. An innovation in fiber-shaped supercapacitor always relates to a revolution in fiber electrodes. The device itself, nothing more than twisting two fibers together, is rarely cared. It should be kept in mind that figures of capacitances, energy, and power densities highlighted by a spate of papers could possibly blur the fact that the twisting exerts a strong impact on the performance of a supercapacitor. As many communications provide longitudinal capacitances to gauge the performance of a fiber-shaped supercapacitor, bypassing the ordinarily used capacitances specified by electrode weight, the screw pitch of the twisting fiber dictates the surface area included within a certain length. Particularly, closely twisting electrodes introduce more surface areas for charge storage than loosely twisting electrodes, which leads to a distinct difference in the resulting capacitance. However, the effect of twisting structure is usually ruled out in most publications pertaining to fiber-shaped supercapacitors.

6.3 Coaxial Fiber-Shaped Supercapacitor

A breakthrough in configuration of fiber-shaped supercapacitor is developing a coaxial structure. The coaxial structure makes the device more integrated than the twisting structure and displays a series of advantages in fiber-shaped supercapacitors. In the following discussion, we will expand the coaxial fiber-shaped supercapacitor from its architecture, fabrication, and, finally, electrochemical performances.

6.3.1 Architecture

Figure 6.7 presents the cross-section view of a fiber-shaped supercapacitor in twisting and coaxial structures. Distinct from the twisting structure where electrodes are placed in parallel, the supercapacitor in coaxial structure exhibits a core–sheath configuration which comprises the inner electrode, outer electrode, electrolyte, and separator. The layout of two electrodes can be geometrically abstracted as a column situated within a cylinder sharing the same center axis. That is what the name "coaxial structure" indicates.

We know that the capacitance of a supercapacitor correlates to the amount of materials. For the electrode with same amount of materials, their cross-section area should be the same. For the cylinder electrode with radius of r, simple mathematics can tell that the radius of the fiber-shaped capacitor device (R_1) would be larger than $2r$. And for the coaxial ones, while the inner electrode is also a cylinder electrode with radius of r, the outer electrode is a hollow circular cylinder with the inner radius slightly larger than r and outer radius R_2, which is also the radius

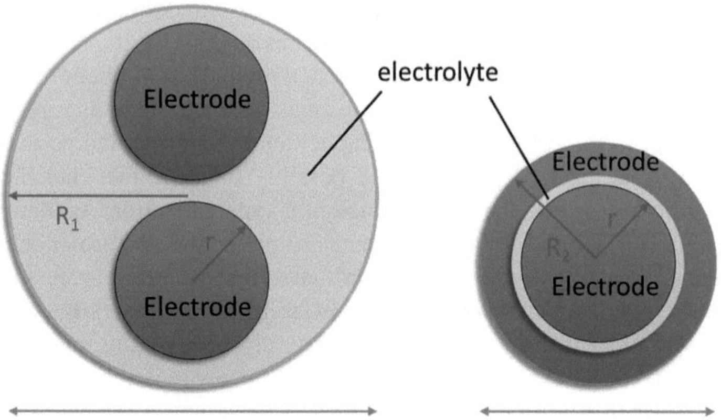

Fig. 6.7 Cross-sectional diagram of fiber-shaped supercapacitor with twisting (*left*) and coaxial (*right*) structure. The diameter of twisting structure would be larger for the same amount of electrode materials

of the whole fiber-shaped device. To obtain the highest specific capacitance, the inner and outer electrodes should have the same amount of active materials. Therefore, the cross-section area of the inner and outer cylinder should be the same. Since the area of circle is πr^2, therefore,

$$\pi r^2 = \pi\left(R_2^2 - r^2\right)$$

R_2 should be 1.414r, while R_1 is 2r.

$$\pi R_1^2 = 4\pi r^2$$
$$\pi R_2^2 = 2\pi r^2$$

And the cross-sectional area of the fiber-shaped supercapacitor with twisting structure would be over $4\pi r^2$, while the coaxial counterpart would be only slightly over $2\pi r^2$, meaning that the coaxial may increase the volume utilization rate by nearly 100 %.

6.3.2 Fabrication

The fabrication of a fiber-shaped supercapacitor in coaxial structure seems more complicated. As depicted in Fig. 6.8, the device is fabricated layer by layer, from inside to outside, in sequence of inner electrode, electrolyte and separator, and outer electrolyte.

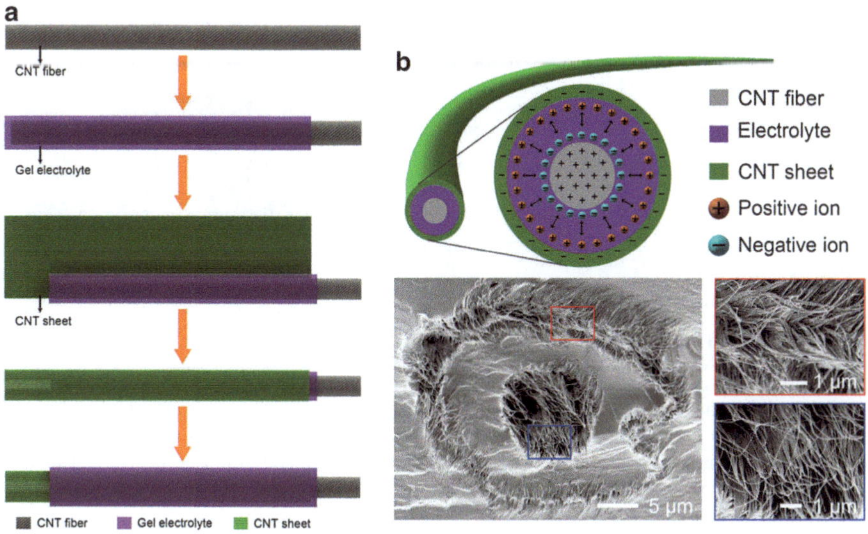

Fig. 6.8 **a** Fabrication procedure of coaxial fiber-shaped supercapacitor by a wrapping method. **b** Structure of coaxial fiber-shaped supercapacitor with carbon nanotube sheet as outside (Reproduced from Ref. [43] by permission of John Wiley & Sons Ltd.)

6.3.2.1 Inner Electrode Materials

The inner electrode has the same requirements on material and structure in line with that in twisting structure: high surface area, conductivity, mechanical strength, flexibility, and chemical stability. Generally, materials like nanowire-modified plastic wire, CNT, graphene, or metal wire with porous carbon are preferred [43–45].

Nanowires were popular in electrochemical applications, since their large specific area and good conductivity benefit the charge storage and transfer. For example, ZnO nanowires have been used to modify a fiber electrode (Fig. 6.9a). To prepare ZnO nanowires, a layer of Au was first sputtered on a plastic wire; then, precursors like zinc nitrate hexahydrate and hexamethylenetetramine (HMTA) were used to prepare nanowires [44]. Moreover, TiO_2-modified Ti wire also has been applied in coaxial supercapacitors. Though the surface area of the metal oxide-modified fiber does not exhibit obvious advantage over the conventional carbon nanomaterials, their properties pertaining to piezoelectric effect and photovoltaic effect provide opportunities to integrate the coaxial supercapacitors with other devices like generators and solar cells.

6.3.2.2 Electrolyte

Electrolyte plays a pivotal role in devices. Having high ionic mobility, solubility, and wettability, liquid electrolyte, nevertheless, is annoyed by problems such as

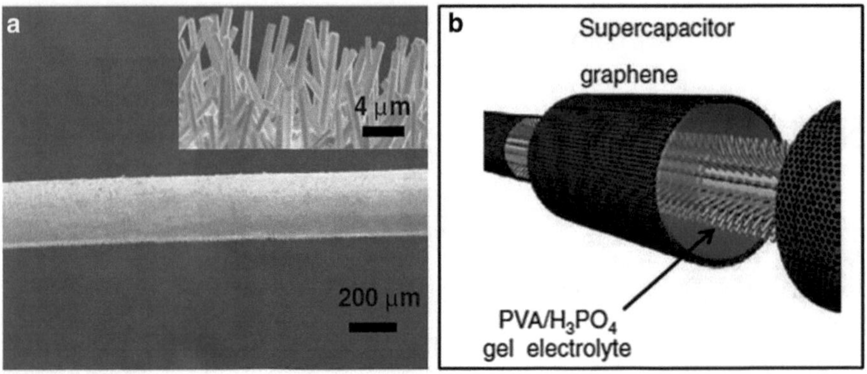

Fig. 6.9 **a** Au-coated plastic wire covered with ZnO nanowire arrays. The inserted SEM image shows that ZnO nanowire arrays were grown along the radial direction. **b** Schematic of supercapacitor part using graphene and gel electrolyte (Reproduced from Ref. [44] by permission of John Wiley & Sons Ltd.)

leakage and corrosion. These problems loom large in fiber-shaped devices due to difficulty on encapsulation. Solid or gel electrolyte can work as ion conductor and separator, offer a better temperature adaptability, circumvent leakage problem, and simplify the fabrication process. Generally, the gel electrolyte is prepared by introducing nonvolatile acid (like H_2SO_4 or H_3PO_4) as ionic conductor into a stable hydrophilic polymer gel such as PVA. A typical route is dissolving PVA into distilled water at 90 °C for 2 h, then cooling down to room temperature, and adding H_3PO_4 aqueous solution. An important effect derived from the electrolyte is that the components in the electrolyte dictate the potential window of the device. Since the water in PVA-H_3PO_4 gel electrolyte delimits the upper limit of the voltage, 1.23 V, the decomposition potential of water, in many cases, the charge voltage is set below 1 V to avoid unwanted side reactions. The voltage can be elevated to 4 V in aprotic electrolyte, making great improvement on both energy density and power density.

6.3.2.3 Outer Electrodes

The outer electrode is the featured part of a coaxial fiber-shaped supercapacitor. The outer electrode, as its name suggests, is placed outmost separated with the inner electrode by electrolyte. The outer electrode can be prepared by directly wrapping afilm-like self-standing material such as graphene, CNT film, and CNT sheet [43, 45–47]. A defect-free large graphene sheet can be synthesized from thermal chemical vapor deposition which was transferred and wrapped onto the device to work as conductive outer electrode [44]. Disordered CNT film was prepared through either filter or coating, then transferred from the substrate, and wrapped onto the device [46]. Due to their structure and surface morphology, CNT film provided large surface area than the graphene sheet, delivering higher specific capacitance. A more easy-made material is the CNT sheet, an anisotropic film where CNTs are oriented. It was directly drawn from the spinnable CNT array [43]. Compared with the disordered CNT film, the CNT sheet exhibits many merits: high conductivity along the alignment, continuous preparation, and tunable thickness beneficial for electrolyte impregnation.

6.3.3 Performances of Coaxial Fiber-Shaped Supercapacitor

6.3.3.1 Electrochemical Property

A prototype of coaxial supercapacitor was first made with ZnO nanowire as the inner electrode and graphene sheet as the outer electrode and gel electrolyte (Fig. 6.8b) [44]. The capacitance of this coaxial supercapacitor was 0.4 mF cm^{-2} at a scan rate of 100 mV s^{-1}. The capacitance is specified to the effective surface area of the inner electrode that was covered with graphene. Considering the one-dimensional configuration of the device, the longitudinal specific capacitance, the capacitance per unit length, is introduced to gauge the coaxial supercapacitor. The as-prepared supercapacitor has a specific capacitance of 0.025 mF cm^{-1}. However, it should be noticed that the longitudinal specific capacitance is convincing provided that all the devices have the same configuration in the same research. The index became less reliable when generalized to different devices. Factors exerting a great impact on the final performances as electrode diameter and helix angle are obscured when indexed by the capacitance lengthwise. So it is unfair to gauge and compare the performances of the twisting and the coaxial supercapacitors. On this account, the longitudinal specific capacitance is not recommended in fundamental research. Admittedly, the capacitance of the rudimentary supercapacitor was not satisfying. Since the double-layer capacitance relies on the surface area accessible for electrolyte, the graphene sheet, whose surface area is dwarfed by the CNT film, was ascribable to the inferior performance. The supercapacitor using CNT film showed a higher specific capacitance of 80 F g^{-1}, which was specified to the weight of CNT only and 11 F g^{-1} specified to two electrodes [46].

In the CNT film, CNT is prone to random aggregation which generates large resistance since electrons have to transfer across massive boundaries in the network. The superiority of CNTs, unfortunately, is fading. The high internal resistance is unfavorable to the energy uptake and delivery. Aligned CNT assemblies including CNT fibers and CNT sheets which inherit merits of individual CNTs became a reliable resolution. Figure 6.8b displays a coaxial supercapacitor that employed CNT fiber as the inner electrode and CNT sheet as the outer electrode. The two electrodes were equal in weight. The electrochemical behavior of this device is representative and exemplifies the merits of coaxial supercapacitor, which is exhibited in Fig. 6.10 and will be discussed in detail. Figure 6.10 displays typical characterizations conducted to gauge the electrochemical performance of a supercapacitor, including galvanostatic charge–discharge profile, rate performance,

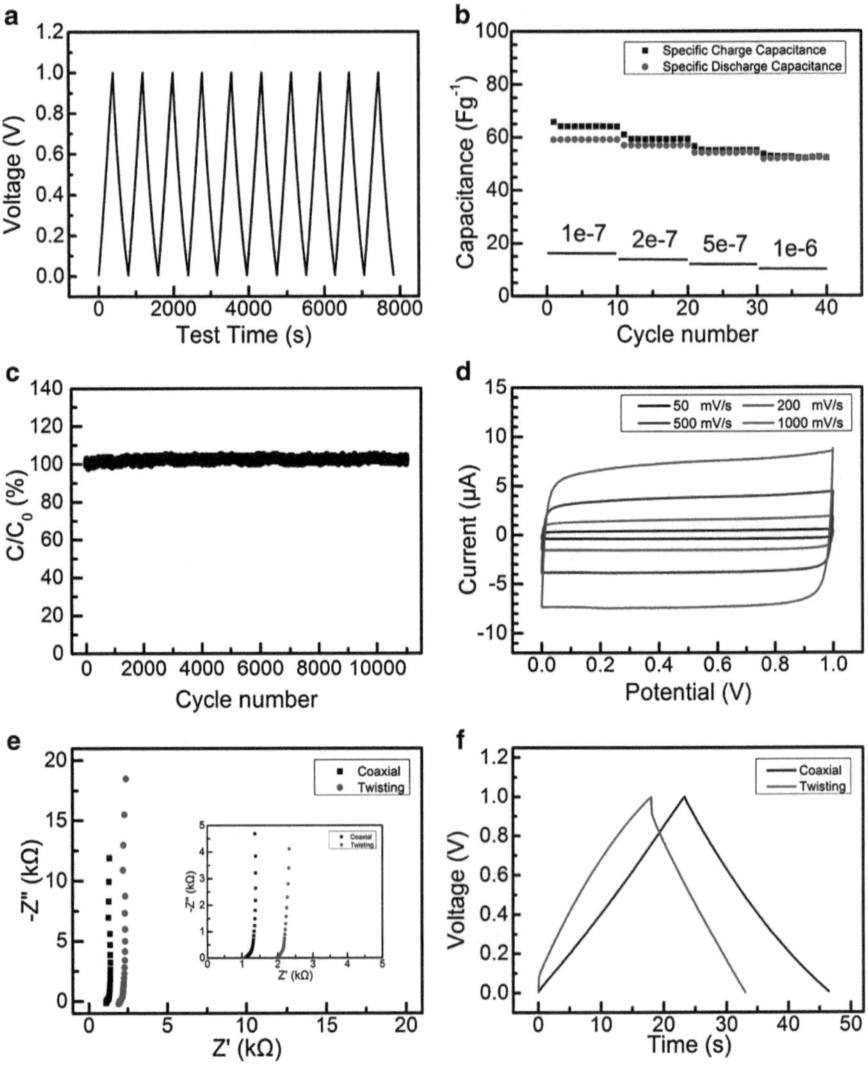

Fig. 6.10 Electrochemical characterization of the coaxial fiber-shaped supercapacitor. **a** Galvanostatic charge–discharge profile at $1.00?\times?10^{-7}$ A. **b** Cycle performance with the increasing electric current. **c** Dependence of capacitance on cycle number at $1.00?\times?10^{-5}$ A. **d** *I–V* curves with the increasing scan rate. **e** Nyquist plots of fiber-shaped supercapacitors with both coaxial (*square dots*) and twisting (*triangular dots*) structures. **f** Galvanostatic charge–discharge curves for coaxial (*solid line*) and twisting (*dotted line*) supercapacitors at 1 μA (Reproduced from Ref. [43] by permission of John Wiley & Sons Ltd.)

cyclic performance, cyclic voltammogram, and electrochemical impedance spectroscopy. Typical galvanostatic charge–discharge profiles do not show any voltage spurts, an upsurge, and drop in voltage arising out of the internal resistance, during charge and discharge processes, which suggests the coaxial structure has smaller resistance than twisting structure (Fig. 6.10a, f).

The rate performance in Fig. 6.10b indicates the kinetics in charge and discharge processes. The capacitance is independent from the current in theory, which, however, is often affected by the kinetics of charge and discharge. It is often the case that the capacitance endures a slump at high current suggesting the charge transfer is impeded by the internal resistances. In this case, the capacitance was maintained by 88 % even when the current was increased by an order of magnitude, which backs up the assertion that the coaxial structure is favorable for kinetics of charge and discharge. The cyclic stability, which is highlighted in double-layer capacitors, is presented in Fig. 6.10c. No discernible decay in capacitance was observed in 11,000 cycles, indicating no failure and deterioration in electrode structure arose after thousands of draining and recharging. Figure 6.12 displays a typical cyclic voltammogram of a double-layer capacitor that is featured as a rectangular shape at different scan rates. The direct evidence that confirms the lower internal resistance of the coaxial device is given by the Nyquist plot from electrochemical impedance spectroscopy, which depicts the response behavior to alternate current with different frequencies (Fig. 6.10e). The interpretation of the Nyquist plot can be complicated involving tedious mathematical calculations and equivalent circuit analysis. Here, we present a sketchy understanding that the intercept with horizontal axis represents the internal resistance combining resistance of the solution, separator, and electrode materials, which is impervious to the frequencies. Obviously, the internal resistance of the supercapacitor in coaxial structure was smaller than the twisting structure. The coaxial supercapacitor with aligned CNT sheet as the outer electrode delivered a specific capacitance of 59 F g^{-1}, based on the individual electrode, in stark contrast to its parallel counterpart, 4.5 F g^{-1}. The advancement of the capacitance should be attributed to the coaxial structure where the outer electrode introduced abundant surface area accessible for electrolyte.

6.3.3.2 Mechanical Property

Apart from higher capacitance and smaller internal resistance, the coaxial structure also exhibits superiorities in mechanical properties. For example, coaxial structure is more integrated than the twisting structure. The electrodes in twisting structure are apt to slip off and separated apart to make the device disassembled, which is aggravated under deformation. The coaxial structure, fortunately, is immune from this concern because two electrodes are unified by gel electrolyte. Hence, the coaxial supercapacitor was more endurable to deformation and retained its intactness, and it showed good flexibility that enabled itself to be bent without degradation in performance (Fig. 6.11a, b). Furthermore, the infiltration of gel electrolyte, which buffers the internal stress when deformed, even endowed the supercapacitor with certain stretchability. As shown in Fig. 6.11c, d, the coaxial supercapacitor afforded to stretch by 20 % for 80 times. In light of its potential application in wearable devices, the coaxial structure has better geometrical regularity, providing better ability for weaving. The stretchability can be enhanced based on the coaxial structure, as discussed in the next section.

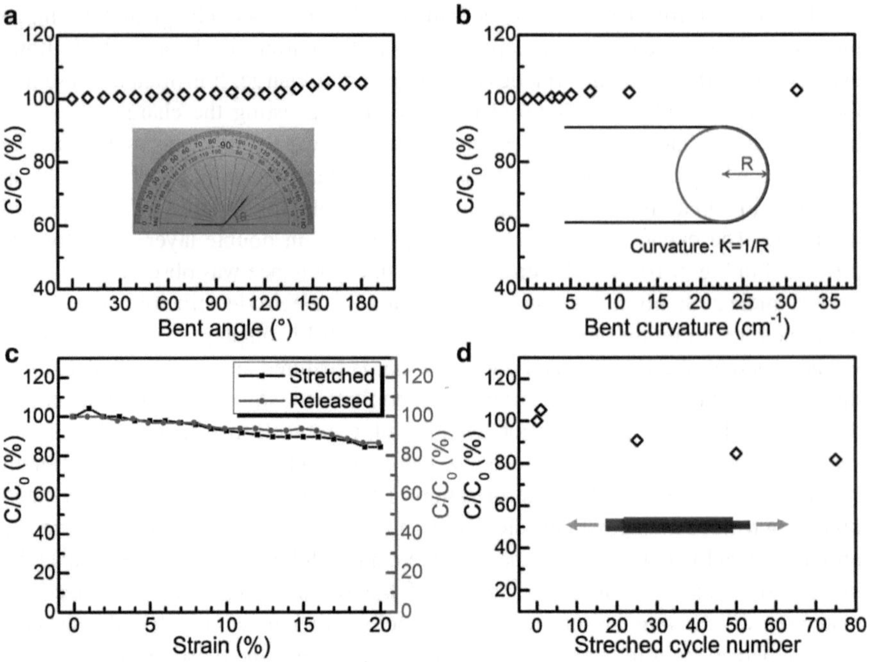

Fig. 6.11 Evolution of specific capacitance during bending (**a** and **b**) and stretching (**c** and **d**) (Reproduced from Ref. [43] by permission of John Wiley & Sons Ltd.)

6.4 Multifunctional Fiber-Shaped Supercapacitors

Inspired by the wearable energy devices, fiber-shaped supercapacitors have witnessed an upsurge in interest. As we discussed previously, various fiber-shaped supercapacitors are developed with the diversification of fiber electrodes. Encouraging progress has been achieved and paints an exciting picture for the application of fiber-shaped supercapacitors. However, some researchers doubt the prospect that the cheer in lab will translate into the success in market because the current fiber-shaped supercapacitors are still wrestling with many problems hiding behind the arresting data highlighted in papers. One breakthrough is extending the functions of the fiber-shaped supercapacitors making them more adaptable and compatible to practical applications. Several proof-of-concept attempts are discussed below.

6.4.1 *Stretchable Supercapacitor*

Imparting stretchability to a fiber-shaped supercapacitor seems a daunting challenge at first glance because fiber electrodes like metal wire, CNT fiber, and graphene fiber are not elastic while the elastic fibers like rubber fibers are not conductive. Yet, on second thought, a stretchable device may not necessitate an elastic electrode. Recently, researchers found that structure engineering, bypassing the contradiction that elasticity is incompatible with conductivity, is a promising approach towards the stretchable supercapacitors. For example, Xu et al. [48] proposed a strategy that introduce an elastic fiber as a substrate and then twine the supercapacitor around the elastic substrate. Technically, a fiber-shaped supercapacitor was fabricated in the beginning by twisting two CNT fibers and coating gel electrolyte. Next, an elastic polymer fiber was pre-stretched to 100 %, and then the fiber-shaped supercapacitor was transferred onto the stretched substrate. When the substrate was released, the affixed supercapacitor would wrinkle rendering the device stretchable up to 100 % (Fig. 6.12a). The stability in deformation and electrochemical performances was inspected by cyclic voltammetry and galvanostatic charge–discharge tests. As displayed in Fig. 6.12b, the cyclic voltammogram retained the featured rectangular shape after 20 cycles of deep stretching. Afterwards, the specific capacitance slightly fluctuated within 10,000 charge–discharge cycles suggesting the stretchable supercapacitor did not degrade after deformation.

The above mentioned device inherited the energy storage capability from the affixed fiber-shaped supercapacitor while realized its stretchability through the auxiliary elastic substrate. Strictly speaking, the device was assembled from discrete units, and the supercapacitor itself did not live up to the conception of stretchability. Moreover, the device was prone to disassemble under repeated deformation. In light of these vulnerabilities, a stretchable supercapacitor based

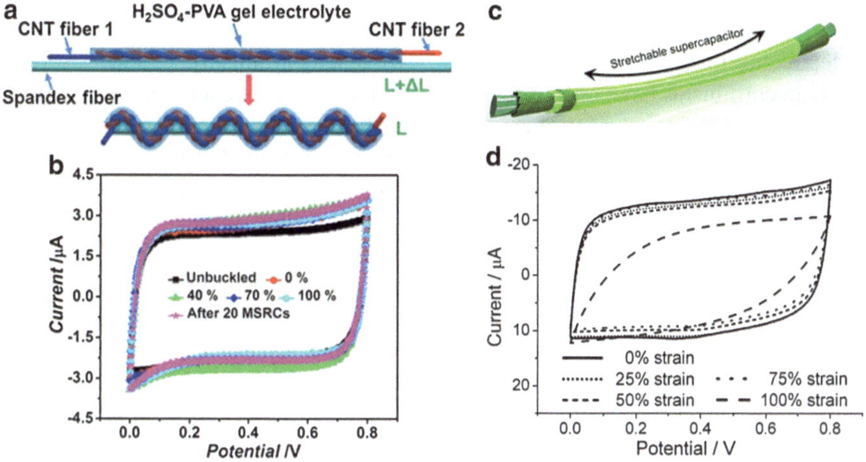

Fig. 6.12 **a** The structure of a fiber-shaped supercapacitor. **b** CV curves of stretchable fiber-shaped supercapacitor at different strains. **c** Schematic illustration to a highly stretchable fiber-shaped supercapacitor in a coaxial structure. **d** CV curves of the fiber-shaped supercapacitor with increasing strains from 0 to 100 % (Reproduced from Refs. [48] and [47] and by permission of John Wiley & Sons Ltd.)

on the coaxial structure (Sect. 6.3) was created [47]. Similarly, a rubbery fiber was used as an elastic substrate. Then, aligned CNT sheets were wound around endowing the rubbery fiber with conductivity on the surface, which is eligible as an electrode. After infiltrating gel electrolyte, another CNT sheet was wound outside followed by infiltrating gel electrolyte again. The produced fiber-shaped supercapacitor was in a coaxial structure as displayed in Fig. 6.12c. The CNT sheets played a pivotal role in the elastic electrode. The surface of the rubbery fiber was paved with layers of CNT sheets, where the oriented CNTs were angled with the substrate, forming an effective conducting network even when the fiber is stretched. In addition, the infiltrated gel electrolyte acted as a binder that incorporated with CNTs. The fiber-shaped supercapacitor was more integrated and afforded an elongation of 75 % without degradation in performance. Figure 6.12d reveals that cyclic voltammogram is distorted when stretched to 100 %, which was ascribed to the severe polarization arising from the breakage in conducting network. Specifically, 95 % of capacitance was left after stretching by 100 % for 100 cycles. The stretched supercapacitor was reliable and can undergo 1,000 cycles of charge and discharge with 90 % of capacitance provided at last.

Incorporating CNT sheets with elastic fiber is a viable approach to prepare an elastic electrode, which can be generalized to other energy devices like solar cells (Chap. 3) and lithium batteries (Chap. 7).

6.4.2 Chromatic Supercapacitor

Chromatism, an academic expression of "color change," is rarely connected with supercapacitors. The chromatism phenomena induced by electric field (electrochromism), magnetic field (magnetochromism), and solvent (solvatochromism) have been intensively studied over the years, but their applications are seldom stretched to energy devices. The chromatic supercapacitor bridges the gap between the chromatism and energy device. Benefited from the coincidence that polyaniline (PANI) exhibits both electrochromism and pseudocapacitance, a chromatic supercapacitor can be materialized. Polyaniline displays different colors at different oxidation states, which corresponds to different charge states in a supercapacitor. Therefore, using polyaniline can on one hand introduce pseudocapacitance and on the other hand endow the supercapacitor with electrochromism. Chen et al. reported a chromatic supercapacitor prepared by electropolymerizing polyaniline onto aligned CNT sheets paved on polydimethylsiloxane (PDMS) substrate as electrodes [49]. The supercapacitor can reversibly change colors in response to different charge states, as polyaniline switches its oxidation states among leucoemeraldine (yellow), the completely reduced state; pernigraniline (blue), the completely oxidized state; and emeraldine (green), the mediate state. To elucidate the chromatism mechanism, a symmetric supercapacitor was fabricated as a demonstration. The supercapacitor was charged between 0 and 1 V. Considering one electrode only, which is defined as positive electrode, when the

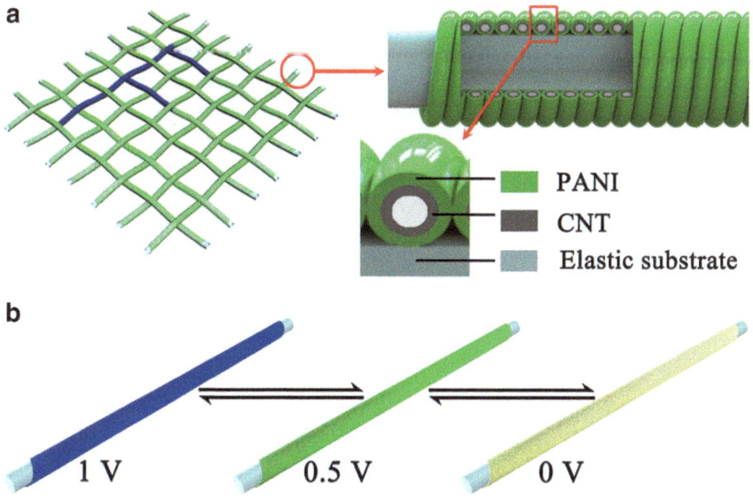

Fig. 6.13 Schematic illustration to the chromatic fiber-shaped supercapacitor. **a** Fabric displayer woven from fiber-shaped supercapacitors and the structure of fiber-shaped supercapacitors. **b** Chromatic behavior at different voltages (Reproduced from Ref. [49] by permission of John Wiley & Sons Ltd.)

supercapacitor was charged to 1 V, the positive electrode was in fully oxidized state and appeared blue; at 0.5 V, it was partially protonated and showed green. When the supercapacitor was drained to 0 V, the electrode turned light green in a slightly oxidized state. If the connection reversed, at voltage of −0.5 V and −1 V, the polyaniline was fully reduced and electrolyte changed to yellow (Fig. 6.13b). In this case, the electrode color is sensitive to the voltage charged, which is useful to monitor the energy storage in the supercapacitor. Furthermore, when a fiber substrate is applied, the chromatic supercapacitor can be fabricated in wire shape and can be woven into textiles that can act as a displayer (Fig. 6.14).

6.4.3 Supercapacitor Textile

The integration of high flexibility and scalable fabrication of wire capacitors promised highly potential application in garment devices, biomedical and antimicrobial textiles, and personal electronics. Today's common garments, such as T-shirts, jeans, leggings, dress shirts, and tank tops, primarily consist of single layers of fabric cut and sewn together. Outerwear garments, such as suit jackets, winter coats, and some sweaters, are made of multiple layers of fabric to achieve a particular shape and/or warmth. However, when manufacturing (weaving or knitting) a textile, they are typically fabricated as single sheets, with their thickness, elasticity, and density determined by their method of construction, structure, and yarn thickness. This presents a unique problem to sandwiched battery and

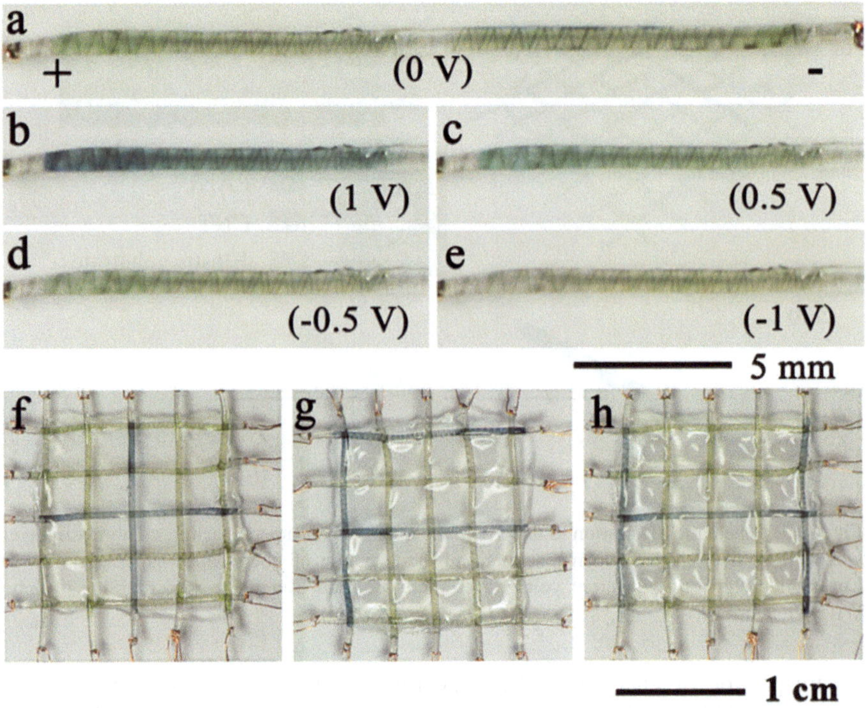

Fig. 6.14 Chromatic behavior during charge–discharge process. **a** Photograph of a chromatic fiber-shaped supercapacitor at 0 V. **b**, **c**, **d**, and **e** Photographs of the positive electrode at 1, 0.5, −0.5, and −1 V, respectively. **f–h** Fabric displayer woven from fiber-shaped supercapacitors (Reproduced from Ref. [49] by permission of John Wiley & Sons Ltd.)

supercapacitor devices, typically comprised of thin layers of current collectors, electrode materials, and a separator. Sandwiched devices would only be optimal for outerwear applications that have sufficient layering and thickness to accommodate many of the proposed devices. Therefore, much of the work conducted above will face manufacturing challenges.

Gorgutsa et al. initially reported a fiber capacitor with strands of their yarns being woven into a full fabric for capacitive touch sensing. This fiber capacitor stored energy on par with small ceramic capacitors ideal for electronic applications like tough sensing, but not larger-scale energy storage. Figure 6.15a shows the great flexibility of a coaxial supercapacitor, enough to make a knot, while Fig. 6.15b further shows a co-woven cloth using multiple cotton yarns and two intact coaxial fibers. The fibers are flexible enough to shuttle back and forth without fracture, as demonstrated by the optical microscopy image of the co-woven cloth. Previously, multiple short CNT/PANI fibers were woven into a piece of cloth, but no electrochemical property was evaluated. Two individual 40 cm long coaxial RGO/CNT@CMC fibers were used as anode and cathode to interweave a cloth supercapacitor. Short circuit of the fiber electrodes was avoided at the crossing of

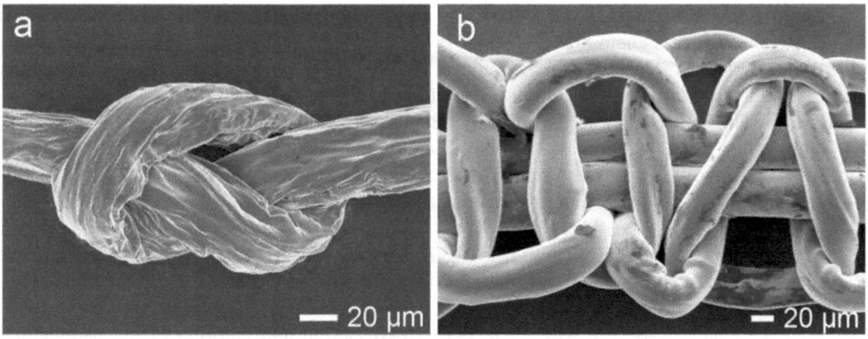

Fig. 6.15 Knotted (**a**) and woven (**b**) coaxial fiber-shaped supercapacitors (Reproduced from Ref. [43] by permission of John Wiley & Sons Ltd.)

the cloth because of the unique coaxial insulated polyelectrolyte-wrapped structure. The cloth supercapacitor showed a capacitance of 28 mF at the current of 10 mA, higher than 25 mF of commercial supercapacitors. The corresponding curves were unchanged under the bending angle of 180° along the three directions, revealing the excellent bendability of the cloth supercapacitors woven from fibers.

References

1. Yan J, Wang Q, Wei T, Fan Z (2014) Recent advances in design and fabrication of electrochemical supercapacitors with high energy densities. Adv Energy Mater 4(4):5166–5180
2. Fu Y, Cai X, Wu H, Lv Z, Hou S, Peng M, Yu X, Zou D (2012) Fiber supercapacitors utilizing pen ink for flexible/wearable energy storage. Adv Mater 24(42):5713–5718
3. Winter M, Brodd RJ (2005) What are batteries, fuel cells, and supercapacitors? Chem Rev 104 (10):4245–4269
4. Simon P, Gogotsi Y (2008) Materials for electrochemical capacitors. Nat Mater 7(11):845–854
5. Kanan MW, Nocera DG (2008) In situ formation of an oxygen-evolving catalyst in neutral water containing phosphate and Co^{2+}. Science 321(5892):1072–1075
6. Liu C, Li F, Ma L-P, Cheng H-M (2010) Advanced materials for energy storage. Adv Mater 22 (8):E28–E62
7. Jost K, Dion G, Gogotsi Y (2014) Textile energy storage in perspective. J Mater Chem A 2 (28):10776–10787
8. Frackowiak E, Beguin F (2001) Carbon materials for the electrochemical storage of energy in capacitors. Carbon 39(6):937–950
9. Zheng J, Cygan P, Jow T (1995) Hydrous ruthenium oxide as an electrode material for electrochemical capacitors. J Electrochem Soc 142(8):2699–2703
10. Wang K, Wu H, Meng Y, Wei Z (2014) Conducting polymer nanowire arrays for high performance supercapacitors. Small 10(1):14–31
11. Wang K, Meng Q, Zhang Y, Wei Z, Miao M (2013) High-performance two-ply yarn supercapacitors based on carbon nanotubes and polyaniline nanowire arrays. Adv Mater 25 (10):1494–1498
12. Oueiny C, Berlioz S, Perrin F-X (2013) Carbon nanotube–polyaniline composites. Prog Polym Sci 39(4):707–748

13. Naoi K, Simon P (2008) New materials and new configurations for advanced electrochemical capacitors. J Electrochem Soc 17(1):34–37
14. Niu C, Sichel EK, Hoch R, Moy D, Tennent H (1997) High power electrochemical capacitors based on carbon nanotube electrodes. Appl Phys Lett 70(11):1480–1482
15. Frackowiak E, Metenier K, Bertagna V, Beguin F (2000) Supercapacitor electrodes from multiwalled carbon nanotubes. Appl Phys Lett 77(15):2421–2423
16. An KH, Kim WS, Park YS, Moon J-M, Bae DJ, Lim SC, Lee YS, Lee YH (2001) Electrochemical properties of high-power supercapacitors using single-walled carbon nanotube electrodes. Adv Funct Mater 11(5):387–392
17. Xu J, Wang Q, Wang X, Xiang Q, Liang B, Chen D, Shen G (2013) Flexible asymmetric supercapacitors based upon Co_9S_8 nanorod/Co_3O_4@ RuO_2 nanosheet arrays on carbon cloth. ACS Nano 7(6):5453–5462
18. Hu C-C, Chang K-H, Lin M-C, Wu Y-T (2006) Design and tailoring of the nanotubular arrayed architecture of hydrous RuO_2 for next generation supercapacitors. Nano Lett 6 (12):2690–2695
19. Choi BG, Chang S-J, Kang H-W, Park CP, Kim HJ, Hong WH, Lee S, Huh YS (2012) High performance of a solid-state flexible asymmetric supercapacitor based on graphene films. Nanoscale 4(16):4983–4988
20. Lu X, Yu M, Wang G, Zhai T, Xie S, Ling Y, Tong Y, Li Y (2013) H-TiO_2@ MnO_2//H-TiO_2@C core-shell nanowires for high performance and flexible asymmetric supercapacitors. Adv Mater 25(2):267–272
21. Liu N, Ma W, Tao J, Zhang X, Su J, Li L, Yang C, Gao Y, Golberg D, Bando Y (2013) Cable-type supercapacitors of three-dimensional cotton thread based multi-grade nanostructures for wearable energy storage. Adv Mater 25(35):4925–4931
22. Chen W, He Y, Li X, Zhou J, Zhang Z, Zhao C, Gong C, Li S, Pan X, Xie E (2013) Facilitated charge transport in ternary interconnected electrodes for flexible supercapacitors with excellent power characteristics. Nanoscale 5(23):11733–11741
23. Jin Y, Chen H, Chen M, Liu N, Li Q (2013) Graphene-patched CNT/MnO_2 nanocomposite papers for the electrode of high-performance flexible asymmetric supercapacitors. ACS Appl Mater Interfaces 5(8):3408–3416
24. Hou Y, Cheng Y, Hobson T, Liu J (2010) Design and synthesis of hierarchical MnO2 nanospheres/carbon nanotubes/conducting polymer ternary composite for high performance electrochemical electrodes. Nano Lett 10(7):2727–2733
25. Lee SW, Kim J, Chen S, Hammond PT, Shao-Horn Y (2010) Carbon nanotube/manganese oxide ultrathin film electrodes for electrochemical capacitors. ACS Nano 4(7):3889–3896
26. Lang X, Hirata A, Fujita T, Chen M (2011) Nanoporous metal/oxide hybrid electrodes for electrochemical supercapacitors. Nat Nanotechnol 6(4):232–236
27. Hu L, Chen W, Xie X, Liu N, Yang Y, Wu H, Yao Y, Pasta M, Alshareef HN, Cui Y (2011) Symmetrical MnO_2-carbon nanotube-textile nanostructures for wearable pseudocapacitors with high mass loading. ACS Nano 5(11):8904–8913
28. Lota K, Khomenko V, Frackowiak E (2004) Capacitance properties of poly (3, 4-ethylenedioxythiophene)/carbon nanotubes composites. J Phys Chem Solids 65(2):295–301
29. Wang K, Zhao P, Zhou X, Wu H, Wei Z (2011) Flexible supercapacitors based on cloth-supported electrodes of conducting polymer nanowire array/SWCNT composites. J Mater Chem 21(41):16373–16378
30. Mondal S, Barai K, Munichandraiah N (2007) High capacitance properties of polyaniline by electrochemical deposition on a porous carbon substrate. Electrochim Acta 52(9):3258–3264
31. Yu Z, Thomas J (2014) Energy storing electrical cables: integrating energy storage and electrical conduction. Adv Mater. doi:10.1002/adma.201400440
32. Bae J, Song MK, Park YJ, Kim JM, Liu M, Wang ZL (2011) Fiber supercapacitors made of nanowire-fiber hybrid structures for wearable/flexible energy storage. Angew Chem Int Ed 50(7):1683–1687

33. Chen T, Qiu L, Yang Z, Cai Z, Ren J, Li H, Lin H, Sun X, Peng H (2012) An integrated "energy wire" for both photoelectric conversion and energy storage. Angew Chem Int Ed 51(48):11977–11980
34. Cai Z, Li L, Ren J, Qiu L, Lin H, Peng H (2013) Flexible, weavable and efficient microsupercapacitor wires based on polyaniline composite fibers incorporated with aligned carbon nanotubes. J Mater Chem A 1(2):258–261
35. Meng Y, Zhao Y, Hu C, Cheng H, Hu Y, Zhang Z, Shi G, Qu L (2013) All-graphene core-sheath microfibers for all-solid-state, stretchable fibriform supercapacitors and wearable electronic textiles. Adv Mater 25(16):2326–2331
36. Lee JA, Shin MK, Kim SH, Cho HU, Spinks GM, Wallace GG, Lima MD, Lepró X, Kozlov ME, Baughman RH, Kim SJ (2013) Ultrafast charge and discharge biscrolled yarn supercapacitors for textiles and microdevices. Nat Commun 4:1970
37. Kou Y, Xu Y, Guo Z, Jiang D (2011) Supercapacitive energy storage and electric power supply using an aza-fused π-conjugated microporous framework. Angew Chem Int Ed 50(37):8753–8757
38. Choi C, Lee JA, Choi AY, Kim YT, Lepró X, Lima MD, Baughman RH, Kim SJ (2014) Flexible supercapacitor made of carbon nanotube yarn with internal pores. Adv Mater 26 (13):2059–2065
39. Ren J, Bai W, Guan G, Zhang Y, Peng H (2013) Flexible and weaveable capacitor wire based on a carbon nanocomposite fiber. Adv Mater 25(41):5965–5970
40. Yu D, Goh K, Wang H, Wei L, Jiang W, Zhang Q, Dai L, Chen Y (2014) Scalable synthesis of hierarchically structured carbon nanotube-graphene fibres for capacitive energy storage. Nat Nanotechnol 9:555–562
41. Wu Z-S, Winter A, Chen L, Sun Y, Turchanin A, Feng X, Müllen K (2012) Three-dimensional nitrogen and boron co-doped graphene for high-performance all-solid-state supercapacitors. Adv Mater 24(37):5130–5135
42. Kou L, Huang T, Zheng B, Han Y, Zhao X, Gopalsamy K, Sun H, Gao C (2014) Coaxial wet-spun yarn supercapacitors for high-energy density and safe wearable electronics. Nat Commun 5:3754
43. Chen X, Qiu L, Ren J, Guan G, Lin H, Zhang Z, Chen P, Wang Y, Peng H (2013) Novel electric double-layer capacitor with a coaxial fiber structure. Adv Mater 25(44):6436–6441
44. Bae J, Park YJ, Lee M, Cha SN, Choi YJ, Lee CS, Kim JM, Wang ZL (2011) Single-fiber-based hybridization of energy converters and storage units using graphene as electrodes. Adv Mater 23(30):3446–3449
45. Harrison D, Qiu F, Fyson J, Xu Y, Evans P, Southee D (2013) A coaxial single fibre supercapacitor for energy storage. Phys Chem Chem Phys 15(29):12215–12219
46. Le VT, Kim H, Ghosh A, Kim J, Chang J, Vu QA, Pham DT, Lee J-H, Kim S-W, Lee YH (2013) Coaxial fiber supercapacitor using all-carbon material electrodes. ACS Nano 7 (7):5940–5947
47. Yang Z, Deng J, Chen X, Ren J, Peng H (2013) A highly stretchable, fiber-shaped supercapacitor. Angew Chem Int Ed 52(50):13453–13457
48. Xu P, Gu T, Cao Z, Wei B, Yu J, Li F, Byun J-H, Lu W, Li Q, Chou T-W (2014) Carbon nanotube fiber based stretchable wire-shaped supercapacitors. Adv Energy Mater 4(3):1300759
49. Chen X, Lin H, Chen P, Guan G, Deng J, Peng H (2014) Smart, stretchable supercapacitors. Adv Mater 26(26):4444–4449

Chapter 7
Fiber-Shaped Lithium Ion Battery

Abstract Commercialized from 1990, lithium ion battery has become a ubiquitous powering supply for a large variety of electronics. Recently, it is facing a challenge and also a chance with the fast developing wearable electronics, e.g., Google Glass, Apple Watch, and Samsung Gear. To embark the "highway" of wearable devices, lithium ion battery should be endowed with some new abilities to be compatible with wearability. Flexibility is therefore highlighted. Furthermore, fiber format tends to achieve a high extent of flexibility and can be woven into textiles by the mature textile technology whose history is as long as that of the human being. Hence, fiber-shaped lithium ion battery has aroused a wide interest. The principle is the same of the planar batteries, which are composed of anode, cathode, and electrolyte/separator. The difference and the pivotal part is to establish a fiber-shaped skeleton, on which the battery components are stacked. This chapter summarizes the recent progress of fiber-shaped lithium ion batteries and their potential wearable applications. In the beginning, the working mechanism, structure, and electrode materials are introduced. Afterwards, several prototypes of flexible lithium ion batteries including bendable, stretchable, and cable-like lithium ion batteries are presented. Then we shift to discuss typical fiber-shaped lithium ion batteries spanning from materials synthesis, structure, and electrochemical performances.

7.1 Overview of Lithium Ion Battery

Batteries have been known for more than two centuries. It is an electrochemical device that can store electricity in the form of chemical energy and reverse the process by converting the chemical energy to electricity. Like its analogues, supercapacitors, a battery is comprised of two electrodes, cathode (positive electrode) and anode (negative electrode), separated by an ionic conductive electrolyte, providing voltage derived from the potential difference between the two electrodes. In accordance to the cyclic performance, they can be classified into two categories: the primary battery and the secondary battery. The secondary battery, also called rechargeable battery, can be reused after draining by recharging. The cyclic performance, the ability to sustain

H. Peng, *Fiber-Shaped Energy Harvesting and Storage Devices*, Nanostructure Science and Technology, DOI 10.1007/978-3-662-45744-3_7

Table 7.1 Summary of the representative commercial rechargeable batteries [1]

Battery	Anode	Cathode	Electrolyte	Voltage (V)	Cycle life
Lithium ion	Graphite	$LiCoO_2$	$LiPF_6$	3.7	>1,000
Lead–acid	Pb	PbO_2	H_2SO_4	2.1	<500
Ni–Cad	Cd	NiOOH	KOH	1.2	2,000
NMH	Intermetallic	NiOOH	KOH	1.2	500–1,000

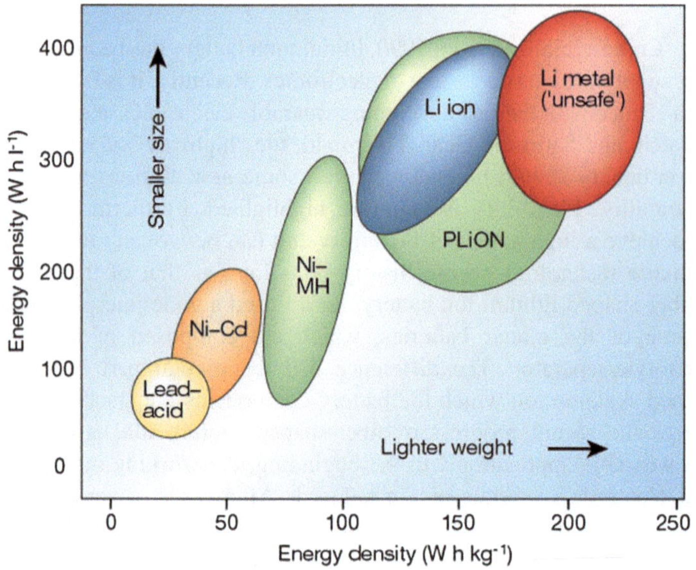

Fig. 7.1 Specific energy densities of the different rechargeable batteries (Reprinted by permission from Nature Publishing Group: Ref. [2], copyright 2001)

repeated charge and discharge, is a critical index to evaluate the performance of a battery. Since the secondary battery is more environment friendly and resource saving, it is preferred by a wide range of applications spanning from onboard power to portable devices. Four types of rechargeable batteries are currently available in the market, as summarized in Table 7.1: lead–acid battery, nickel–cadmium battery, nickel–metal hydride battery, and lithium ion battery [1]. As depicted in Fig. 7.1, among various rechargeable batteries, lithium ion battery stands at the leading position and surpasses its rivals with the highest energy density [2].

7.1.1 General Principle

In 1990, Sony announced a novel high-performance rechargeable battery with $LiCoO_2$ as the cathode-active material and a tailor-made carbonaceous material as an anode. In such batteries, lithium ions are inserted and extracted from electrodes and migrate back and forth in the electrolyte during repeated charge and discharge processes, thus delivering and restoring energy. In subsequence of its emergence, the lithium ion battery successfully penetrated into the market and swept the world in the following years. The popularity of lithium ion battery is benefited from its arresting merits including high energy density, decent cyclic longevity, and no memory effect that annoys its competitors, nickel–cadmium (Ni–Cd) or nickel–hydride (Ni–MH) batteries.

Earlier attempts used lithium metal as the anode due to its high electrochemical capacity (3,860 mAh g^{-1}), very negative standard redox potential (−3.04 V vs. NHE), and low density (0.53 g cm^{-3}), but it was soon aware that the lithium metal will bring about the unwanted parasitic reactions on the lithium surface, thus rendering the battery limited cyclic longevity. More importantly, the lithium metal is prone to form lithium dendrite which causes safety risk of the battery. Considering the negative effect of lithium metal, the anode was then replaced by graphitic carbon for lithium reservoir. It has been clarified that lithium ions can be inserted into the crystal lattice of graphitic carbon forming intercalation compound (LiC_6). The voltage profile presents a stable plateau at ~0.2 V vs. Li/Li^+ when lithiated and delithiated. The lithiated compound, LiC_6, determines that theoretical lithium storage capacity is 372 mAh g^{-1}.

In the case of cathode materials, layered compounds where lithium ions are readily intercalated with maintained structure are preferred, as exemplified by transition metal sulfides and $LiCoO_2$. Titanium disulfide (TiS_2) is one of the most appealing layered dichalcogenides applicable as energy storage materials. When compounded with lithium ions, the lithiated compounds Li_xTiS_2 exhibits no phase transition, exempt from the rearrangement and reconstruction of lattice, which is beneficial for a remarkable reversibility. $LiCoO_2$ has a similar layered structure with TiS_2 but exhibits different lithiation behavior involving several phases. Therefore, only half of the lithium in the compound can be exploited to deliver energy leading to a relatively low capacity in theory of 140 mAh g^{-1}. SONY developed the first lithium ion battery that dominates the market a decade after with $LiCoO_2$ coupling with graphite as cathode and anode [3]. Originated from the pioneer works, lithium ion battery cathodes are diversified under intensive investigations over the years, as presented in Table 7.2 [4].

In general, energy delivery and storage of lithium ion batteries are based on the migration of lithium ions between the two electrodes, which serve as lithium host through "electrochemical intercalation," in charge and discharge process (Fig. 7.2). Exemplified with $LiCoO_2$–graphite system, the main cell reactions are reversible lithium ion insertion and extraction process between two layered compounds. The reaction in cathode can be expressed as

$$LiCoO_2 \leftrightarrows \frac{1}{2}Li^+ + \frac{1}{2}e^- + Li_{0.5}CoO_2 \qquad (7.1)$$

The upper voltage of the delithiation of $LiCoO_2$, the charge process, is limited to 4.2 V (vs. Li/Li^+), which means only half of the lithium stored in the cathode can be

Table 7.2 Properties of the representative cathode materials for lithium ion batteries [4]

Material	Capacity ($mAh\ g^{-1}$)	Density ($g\ mL^{-1}$)	Energy density ($mAh\ mL^{-1}$)	Shape of dis. curve	Safety	Cost
$LiCoO_2$	160	5.05	808	Flat	Fair	High
$LiNiO_2$	220	4.80	1,056	Sloping	Poor	Fair
$LiNi_{0.8}Co_{0.2}O_2$	180	4.85	873	Sloping	Fair	Fair
$LiNi_{0.8}Co_{0.15}Al_{0.05}O_2$	200	4.8	960	Sloping	Fair	Fair
$LiMn_{0.5}Ni_{0.5}O_2$	160	4.70	752	Sloping	Good	Low
$LiMn_{1/3}Ni_{1/3}Co_{1/3}O_2$	200	4.7		Sloping	Good	Low
$LiMn_{0.4}Ni_{0.4}Co_{0.2}O_2$	200	4.7				
$LiMn_2O_4$	110	4.20	462	Flat	Good	Low
$Li_{1.06}Mg_{0.06}Mn_{1.88}O_4$	100	4.20	420	Flat	Good	Low
$LiFePO_4$	160	3.7	592	Flat	Good	Low

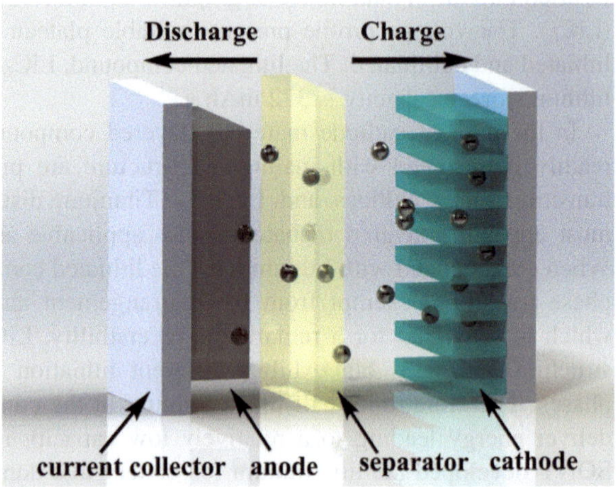

Fig. 7.2 Schematic illustration to a lithium ion battery

extracted [5]. When the battery is overcharged, the parasitic reactions of electrolyte decomposition will prevent the further delithiation. $LiCoO_2$ as lithium source ensures a long shelf life and excellent stability. When fabricated, the battery is in discharged state and should be charged first before use. During charge process, the lithium is extracted from the cathode and migrates towards the anode. The lithiation and delithiation of graphite is expressed as

$$C_6 + Li^+ + e^- \rightleftharpoons LiC_6 \quad (7.2)$$

With the intercalation of lithium ions, the graphite experiences several stages of phase transition spanning from LiC_{24}, LiC_{27}, and LiC_{12} to the final product LiC_6 [6]. In the first charge cycle in polar aprotic electrolyte, a passivation film is apt to form at the surface of the anode, derived from the decomposition of electrolyte, which on one hand protects the graphite from deterioration and on the other hand causes initial irreversible capacity. Hence, lithium ion battery has to keep the lithium stored in electrode in excess, in order to satisfy the consumption of the passivation films on the graphite.

The story for the electrolyte was somehow more straightforward as aqueous solutions are excluded due to the high reactivity of water with lithium metal and also its limited temperature window (0–100 °C) and decomposition potential (1.23 V vs. NHE). Electrolyte of a lithium salt in aprotic solvents with wide electrochemical window and good compatibility is found effective in a lithium ion battery. In the following sections, we will detail our discussion in the cathode, anode, and electrolyte.

7.1.2 Cathodes

The lithium ion battery generally outputs a voltage of more than 3.5 V with a pair of cathode material and carbonaceous anode material, in which the lithium ion reversibly inserts and extracts. Such electrochemical reaction proceeds at a potential of 4 V vs. Li/Li^+ electrode for cathode and ~0 V for anode. Since the product of its output voltage and capacity dictate the energy of a battery, a high voltage and capacity along with light weight promise a high energy density, which is badly desired and vigorously pursued. Therefore, the cathode materials play a pivotal role to the energy density of the battery. Table 7.2 summarizes the structures and electrochemical properties of representative cathode materials. The current state-of-the-art for cathode electrodes includes $LiCoO_2$, $LiNi_{1/3}Mn_{1/3}Co_{1/3}O_2$, $LiMn_2O_4$, as well as the recently discovered polyanionic compounds like $LiFePO_4$ and $LiFeSO_4$, which are poised to play a major role in the near future.

7.1.2.1 $LiCoO_2$ with a Layered Structure

$LiCoO_2$, which is the most commonly used cathode material, has a theoretical capacity of 274 mAh g^{-1} provided that all the lithium is extracted. But unfortunately, only half of lithium stored in $LiCoO_2$ is cyclically available corresponding to the charge ceiling of 4.2 V; otherwise the structure inclines to collapse due to a significant electrostatic repulsion between the transition metal polyhedral. Hence, the reversible capacity is typically limited to around 140 mAh g^{-1}. To overcome this problem, strategies have been devised that partially displace a fraction of the cobalt in this phase with nickel or manganese so as to preserve the structure against decomposition at high potential. The stoichiometric formula of the resulting product is $LiCo_{1/3}Ni_{1/3}Mn_{1/3}O_2$ which promises a reversible capacity of 190 mAh g^{-1} [7]. The capacity can be further increased by synthesizing lithium-rich layered oxides by embedding inactive Li_2MnO_3 component within layered $LiMO_2$. The as-prepared $xLi_2MnO_3 \cdot (1\text{-}x)LiMn_{0.5}Ni_{0.5}O_2$ composite displays a capacity of 250 mAh g^{-1} [8].

7.1.2.2 $LiMn_2O_4$ with a Spinel Structure

In contrast, the structure of spinel $LiMn_2O_4$ proves to be much more robust against the removal of lithium, capable of reversibly removing and restoring utmost 95 % of the theoretical capacity. The combination of the spinel 3D lattice with chemical stability of the Mn^{+3}/Mn^{+4} couple promises good safety and high power capability for a cathode material. Manganese is also a relatively economical material. However, this material also endures stability problems such as self-discharge when used in conventional liquid electrolytes where the trace amounts of HF will corrode the material, especially at elevated temperatures. $LiMn_2O_4$ spinel is in fact the fastest cathode material due to its unique 3D structure, which allows the multi-directional diffusion of Li ions, which delivers a capacity of 80 mAh g^{-1} at high drain rates [9, 10].

7.1.2.3 $LiFePO_4$ with an Olivine Structure

$LiFePO_4$, which has an olivine structure, has several advantages, including environmental friendliness, low cost, no toxicity, chemical stability, high capacity, and thermal stability. The discharge potential of pure $LiFePO_4$ is around 3.45 V (vs. Li/Li^+), and the theoretical capacity is 170 mAh g^{-1}. However, the insulating nature (10^{-9} S cm^{-1}) of $LiFePO_4$, leading to slow lithium ion diffusion ($<10^{-14}$ cm^2 s^{-1}), high impedance, and inferior rate capability, is the crux impeding them towards commercial applications [11]. On this account, in order to improve the electrochemical performance of $LiFePO_4$ cathode, various efforts have been devoted to decreasing the particle size and increasing electrical conductivity. It has been demonstrated that the performances of $LiFePO_4$ can be improved by structural or morphological modifications, including mechanochemical activation, coating with conducting additives, and supervalent metal ion doping [12, 13].

7.1.3 Anodes

Lithium dendrite has plagued all the lithium ion batteries using lithium metal as the anode. When lithium dendrite pierces the separator and causes an internal short circuit, an explosion hazard ensues. In light of the safety vulnerability of lithium metal, finding a suitable substitute becomes a pivotal issue in the research of anodes. Alloys with aluminum, silicon, zinc, etc., have been under investigation as alternative anode materials since Dey in 1971 demonstrated the possibility of electrochemical formation of lithium alloys in organic electrolytes [14]. Nevertheless, the lithium storage capacity of alloys degrades quickly after merely several charge–discharge cycles arising from the large volume change which causes internal stress in alloy crystal lattice and thus brings about cracking and pulverization of alloy particles. On other side, Sony Corporation succeeded in discovering the highly reversible, low-voltage carbonaceous material as anode and commercialized the C–$LiCoO_2$ rocking chair cells in 1991. Displacing the lithium metal with lithiated carbon exempts the battery from the growth of irregular dendritic lithium, which promises a better compatibility to applications.

7.1.3.1 Carbonaceous Anodes

As early in the 1800s, the layered carbon was found accessible for various species. The capability for intercalation promoted the application of graphite as ion host in rechargeable batteries. Graphite intercalation compounds have one significant feature: the staging phenomenon, which is characterized by a periodic sequence of intercalant layers (Li^+) between graphite layers. The nth-stage compound consists of intercalant layers arranged between every n graphite layers. The first-stage lithium graphite intercalation compound has the stoichiometry of LiC_6 with the specific capacity of 372 mAh g^{-1}, a theoretical saturated value of lithium storage for graphite under normal pressure [15].

7.1.3.2 Metal Alloy Anodes

Generally, a higher capacity of anode is always beneficial to the battery performance. The advantage can be quantitatively discussed. For a given cathode material, with its capacity (C_C) ranging from 140 to 200 mAh g^{-1}, there is a simple relationship between the total capacity of the lithium ion battery (based on all the active materials) and the anode capacity (C_A):

$$C_{\text{battery}} = \frac{C_c}{1 + \frac{C_c}{C_A}} = \frac{C_A C_C}{C_A + C_C} \left(\text{mAh g}^{-1}\right) \tag{7.3}$$

Equation 7.3 reveals that for a given C_C, the total specific capacity does not follow a linear relation with C_A. Figure 7.3 suggests that a noticeable improvement in lithium ion battery can be achieved provided that the carbonaceous anode is replaced with one having a capacity in the order of 1,000 mAh g^{-1} [16].

The study in alloy anode is driven by the fact that the potential of the anode vs. Li/Li^+ should be close to 0 V. Hence, the electrochemical reaction at the anode site does not necessitate a lithium intercalation process whose capacity is restricted by its crystal structure. Instead, lithium ions can react with metal-forming alloy (Li_xM), which can accommodate and deliver more lithium ions. Numerous metals, such as Si, Al, Sn, Sb, Ge, Pb, and Ag, are able to alloy electrochemically with lithium when polarized to a sufficiently negative potential in electrolyte. The formation and decomposition of Li_xM contribute to the high theoretical capacity. Unfortunately, most high-capacity anodes based on the aforementioned metals are often undermined by their short cycle life arising from the large volume expansion and electrolyte depletion [17].

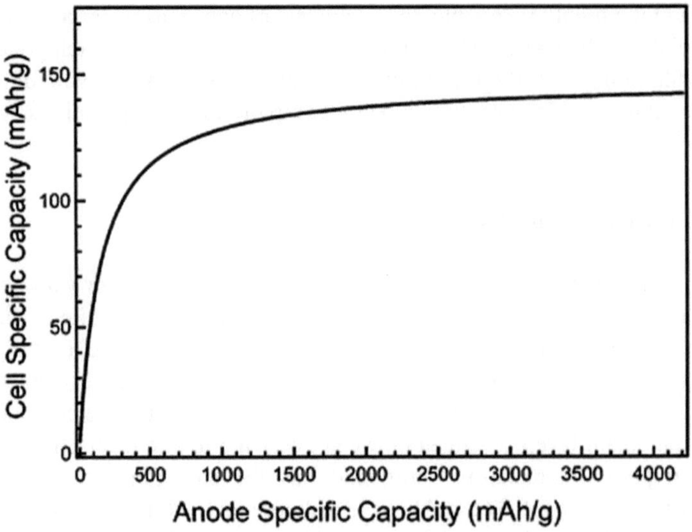

Fig. 7.3 The relation between the total capacity of a lithium ion battery with the anode capacity. The battery comprised of a Si anode and $LiCoO_2$ cathode (the theoretical capacity of $LiCoO_2$ is 147 mAh g^{-1}) (Reproduced from Ref. [16] by permission of The Royal Society of Chemistry)

7.1.4 *Electrolytes*

In lithium ion battery, liquid electrolytes act as ionic conductor to transport lithium ions back and forth between positive and negative electrodes during charge and discharge process. Since the electrodes in lithium ion batteries are the porous composite electrodes, comprising active materials, conducting additives, and a polymer binder, the liquid electrolyte must infiltrate into the porous electrodes and transfer lithium ions smoothly at the interfaces between the liquid and solid phases [18].

Most of the liquid electrolytes used in the commercial lithium ion batteries are the nonaqueous solutions, in which roughly 1 mol L^{-1} of $LiPF_6$ salt is dissolved in the mixture of carbonate solvents selected from cyclic carbonates–ethylene carbonate (EC), and propylene carbonate (PC), and linear carbonates–dimethyl carbonate (DMC), ethyl methyl carbonate (EMC), and diethyl carbonate (DEC). Polymer electrolyte was launched in 1973 by Fenton et al.; its technological importance was appreciated because it can prevent internal short circuit and electrolyte leakage and has a good adaptability to environment. Moreover, the noncombustible reaction products make the battery safer than that with liquid electrolyte.

7.2 Flexible Lithium Ion Batteries

The passion in flexible devices has spread to lithium ion batteries. In recent years, flexible lithium ion batteries emerged to satisfy the ever-growing requirement. In this section, we retrospect the advancement of flexible lithium ion batteries in planar shape and then present an advanced branch of flexible device, cable-type lithium ion battery.

7.2.1 Flexible Lithium Ion Batteries in Planar Shape

The electrodes are the pillar of a battery. Then fabricating a flexible lithium ion battery is reduced to preparing flexible electrodes where active materials are incorporated. Cui et al. has demonstrated a prototype flexible lithium ion battery constructed on free-standing CNT films as current collectors for both anode and cathode. $Li_4Ti_5O_{12}$ (LTO) and $LiCoO_2$ (LCO) were used as anode and cathode materials respectively, which then were paved on the CNT substrates by doctor blade. The two CNT composite films were laminated onto a single sheet of commercial paper that functioned as separator membrane and mechanical support and then sealed in a polydimethylsiloxane (PDMS) and had a thickness of 300 μm (Fig. 7.4a, b). The lithium ion battery inherited the flexibility from the flexible CNT film and paper separator, exhibiting a high energy density of 108 Wh kg^{-1} [19]. The flexible battery relies on the conductive flexible electrodes. Similarly, the CNT film was displaced by a three-dimensional interconnected graphene foam incorporating $Li_4Ti_5O_{12}$ and $LiFePO_4$ as anode and cathode (Fig. 7.4c). Benefited from the highly conductive pathways for electrons and ions provided by the graphene network, the flexible battery can perform at a high current rate of 10 C, giving a capacity of 117 mAh g^{-1}, 88 % of the capacity at a 1 C rate. Ninety-six percent of capacity retained after running for 100 cycles at this high rate [20]. The capacity retention over the high current rate is an important index to gauge the performance of a battery, which relates to the electrical conductivity and ionic mobility inside the battery. Hardly can a flexible battery afford a high rate running because it is often the case that the flexibility sacrifices conductivity. On this account, this flexible battery based on graphene foams enabling a high rate performance is appreciated as a great achievement. The performance of the flexible lithium ion battery can be improved by replacing the active materials and diversifying the nanomorphologies. For example, hierarchical three-dimensional $ZnCo_2O_4$ nanowire arrays were hydrothermally synthesized on a carbon cloth as anode which was paired with $LiCoO_2$/Al cathode and sealed in a flexible plastic package (Fig. 7.4d). The flexible battery outputs a voltage of 3.4 V and a stably going reversible capacity of 1,300 mAh g^{-1} for 40 cycles [21]. These rudimentary flexible batteries, however, still used liquid electrolyte, making them less adaptable for application.

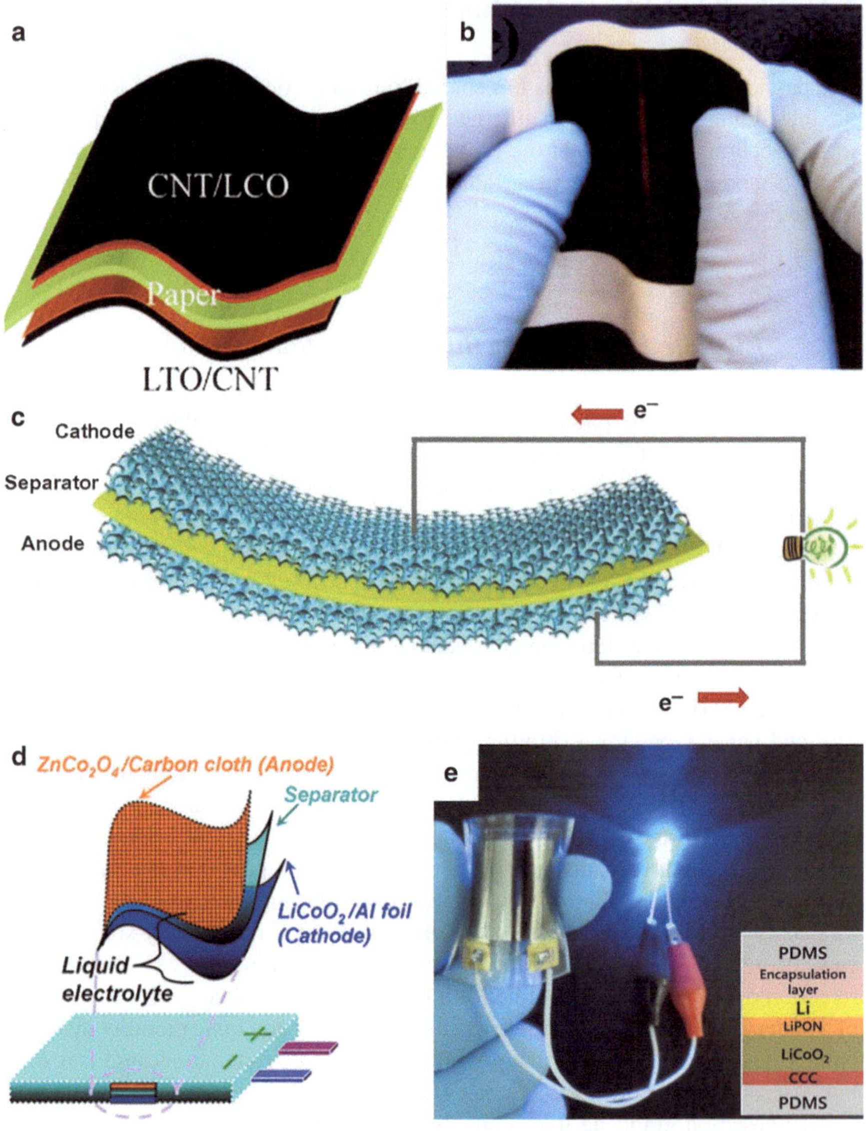

Fig. 7.4 Flexible lithium ion batteries in planar shape. **a** and **b** Structure and photograph of thin-film batteries fabricated from CNT films and paper separator (Reprinted with the permission from Ref. [19]. Copyright 2010 American Chemical Society). **c** Schematic illustration to a flexible battery based on 3D interconnected graphene foams (Reproduced with permission [20]. Copyright 2012, National Academy of Sciences, USA). **d** Hierarchical 3D $ZnCo_2O_4$ nanowire array/carbon cloth and $LiCoO_2$/Al flexible lithium ion battery (Reprinted with the permission from Ref. [21]. Copyright 2012 American Chemical Society). **e** Photograph of a bent lithium ion battery powering a blue LED. The structure of the flexible lithium ion battery is displayed in inset (Reprinted with the permission from Ref. [22]. Copyright 2012 American Chemical Society)

Therefore, solid-state electrolyte like lithium phosphorus oxynitride electrolyte (LiPON) was introduced, which even made the metallic lithium anode compatible to flexibility (Fig. 7.4e) [22].

7.2.2 *Stretchable Lithium Ion Batteries*

As the conception of wearable devices like epidermal sensors gains increasing popularity, the requirement for flexibility is further extended that the device should sustain the tensile deformation under stretch. On this occasion, stretchable lithium ion batteries were developed in response. Here, we will demonstrate two representative prototypes of stretchable lithium ion batteries materialized from two different approaches.

In 2013, Rogers et al. reported a stretchable rechargeable battery device constructed on a silicone elastomer with segmented layouts and iterative interconnection in self-similar geometries [23]. Figure 7.5a displays the structure of the

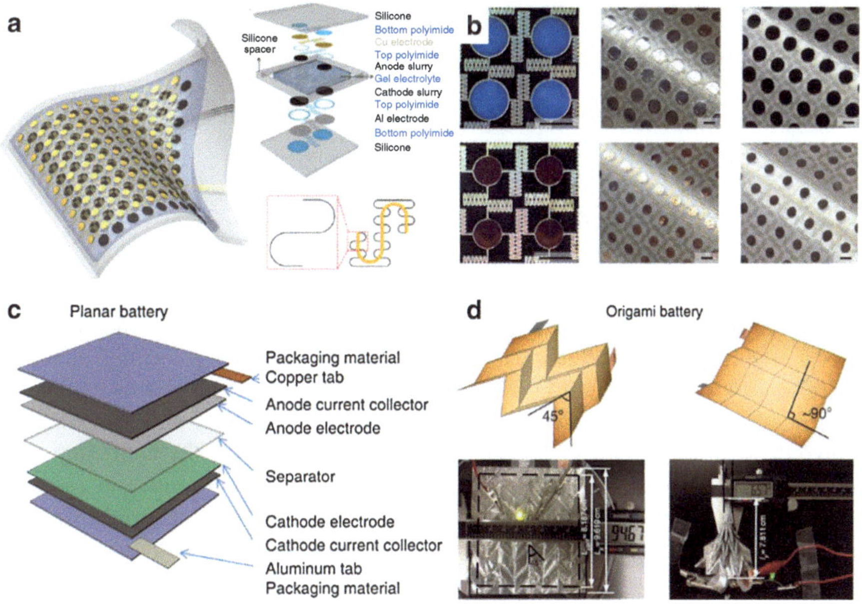

Fig. 7.5 Stretchable lithium ion batteries. **a** Schematic illustration of a completed device, the structure of a unit cell, and self-similar interconnections in serpentine geometries. **b** Optical images of cathode unit array (*up*) and anode unit array (*down*). Scale bars in **b** are 2 mm (Reprinted by permission from Nature Publishing Group: Ref. [23], copyright 2013). **c** Exploded view of the structure of the lithium ion battery. **d** Illustrations and photographs of the origami battery in the planar and completely compressed configurations (Reprinted by permission from Nature Publishing Group Ltd: Ref. [24], copyright 2014)

device with unit cells arrayed on an elastic substrate. Aluminum and copper disks were used as current collectors for cathode and anode where $LiCoO_2$ and $Li_4Ti_5O_{12}$ served as active materials, respectively. The electrode units were connected in a serpentine geometry that makes the device insusceptible to biaxial stretch deformations (Fig. 7.5b). Then the two sheets, the anode unit array and cathode unit array, were laminated and the gel electrolyte was injected. The unit cells were connected in parallel in the device, which output a voltage plateau around 2.35 V. The device can be stretched up to 300 % without degradation in performance. Moreover, the device was soft and compliant, which was adaptable for biological epidermis.

Another strategy for fabricating stretchable battery comes from the origami, the art of paper folding. The origami battery started from the conventional lithium ion battery with CNT papers as current collectors, $LiCoO_2$ and $Li_4Ti_5O_{12}$ as active materials, and liquid electrolyte packaged within an aluminized polyethylene (Fig. 7.5c) [24]. Then the origami was obtained by folding the planar battery in a specific origami pattern called Miura-ori where the battery was divided into many units in identical parallelogram through predefined creases. The Miura-ori pattern made the battery compressible and collapsible. Since the deformability of the origami battery was derived from the prescribed creases, the origami pattern did not sustain large strain. Indicated by an LED, the origami battery exhibited a stable performance during the planar and completely compressed configurations (Fig. 7.5d).

7.2.3 *Cable-Type Lithium Ion Battery*

In 2012, Kim et al. introduced a cable-type lithium ion battery which exhibited good mechanical flexibility and discharge capacity [25]. This new type of battery has a sophisticated structure. Cu wires deposited with Ni–Sn active materials were first twisted to form a bundle, four of which were then used to fabricate a hollow-structured anode. The anode was subsequently wound with the modified poly(ethylene terephthalate) (PET) nonwoven separator and the aluminum wire as the cathode current collector, followed by coating $LiCoO_2$ slurry to form the cathode. The obtained electrode assembly was sealed into a heat-shrinkable tube, and finally the aqueous electrolyte was injected into the tube to get the battery. The fabrication process and the structure of the battery are shown in Fig. 7.6a, b. This cable-type battery delivered a capacity of 1 mAh cm^{-1} and a voltage plateau at about 3.5 V. The hollow anode facilitated the electrolyte accessibility to all components of the battery. Besides, this battery is mechanically flexible. Its discharge characteristics can be well maintained in the bending process, and it can power a red LED screen under twisting (Fig. 7.6c).

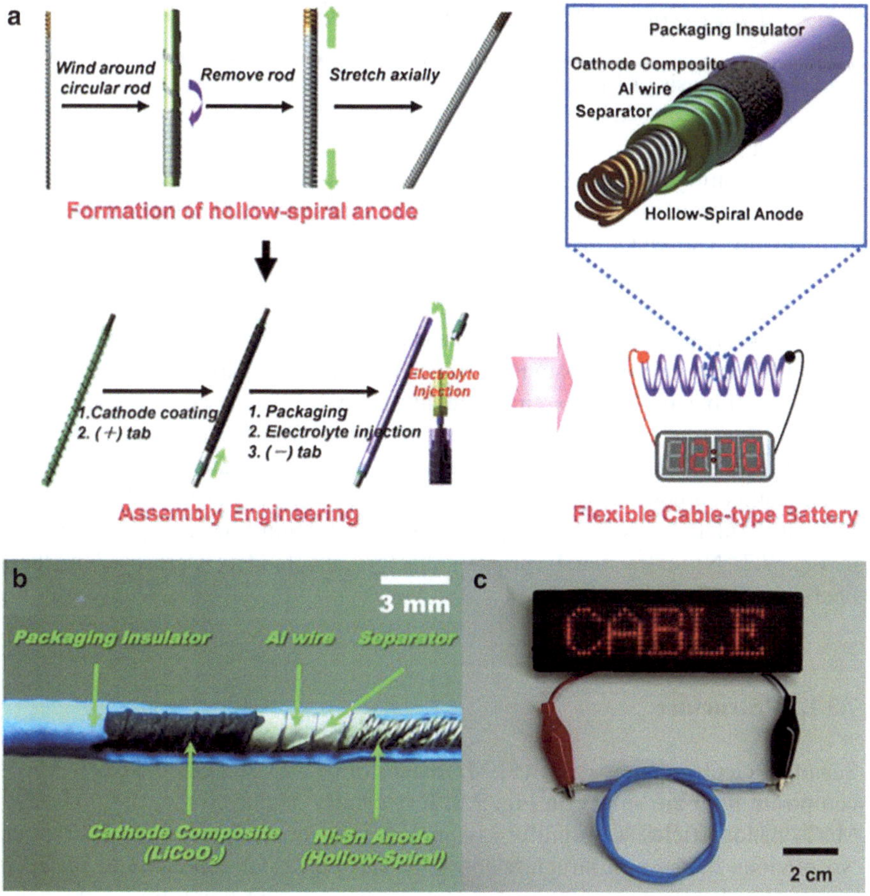

Fig. 7.6 Cable-like lithium ion battery. **a** Schematic illustration to the fabrication and structure of the battery. **b** Photograph of the side view of a cable-type lithium ion battery. **c** Photograph of a 25 cm cable-type battery powering a red LED screen (Reproduced from Ref. [25] by permission of John Wiley & Sons Ltd)

7.3 Fiber-Shaped Lithium Ion Batteries

In this section, we will introduce a classic of fiber-shaped lithium ion batteries based on a novel fiber electrode: aligned carbon nanotube (CNT) fiber. The preparation and property of the carbon nanotube fiber have been discussed in Chap. 2. Here, we mainly concentrate on the fiber-shaped lithium ion batteries employing the CNT fibers, which are incorporated with active materials, as cathodes or anodes. The discussion with several fiber-shaped batteries as examples will be expanded from the material preparations, device structures, and electrochemical performances.

7.3.1 *Fiber-Shaped CNT/MnO_2 Cathode*

7.3.1.1 Synthesis of Materials

In 2012, a fiber-shaped lithium ion battery was created which used aligned CNT fiber compositing MnO_2 as cathode [26]. As described in Chap. 2, the CNT fiber was prepared from the spinnable CNT array with controlled diameters ranging from 20 to 30 μm. The MnO_2 nanoparticles were incorporated to the CNT fiber through electrochemical deposition in an aqueous solution. The content of MnO_2 was controlled by the cycle number of electrochemical deposition, ranging from 0.5 to 8.6 wt%. The high flexibility, mechanical strength, and electrical conductivity of the fiber were impervious to the composition of MnO_2 particles.

As a trailblazing attempt, metallic lithium wire was used as anode where the CNT/MnO_2 composite fiber was intertwisted. The electrolyte consisted of 1 mol L^{-1} of $LiPF_6$ in a mixture solvent of ethylene carbonate, diethyl carbonate, and dimethyl carbonate (weight ratios of 1/1/1) (also called LB303). The schematic illustration of the proof-of-concept fiber-shaped lithium ion battery is displayed in Fig. 7.7. No extra metal current collectors and binders were used in these fiber-shaped batteries.

7.3.1.2 Structure

Scanning electron microscopy (SEM) images of bare CNT fiber and CNT/ MnO_2 composite fiber are shown in Fig. 7.8. It is clearly displayed that the flowerlike MnO_2 nanoparticles were uniformly strewn over the surface of the CNT fiber. No obvious damage in structure integrity and electrical conductivity decay is traced after bending for 100 cycles, suggesting a decent flexibility of the composite fiber.

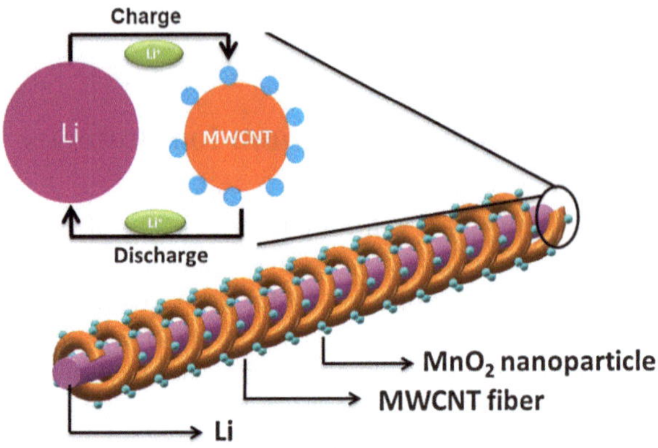

Fig. 7.7 Schematic illustration to a fiber-shaped lithium ion battery with lithium wire as anode and CNT/MnO_2 composite fiber as cathode (Reproduced from Ref. [26] by permission of John Wiley & Sons Ltd)

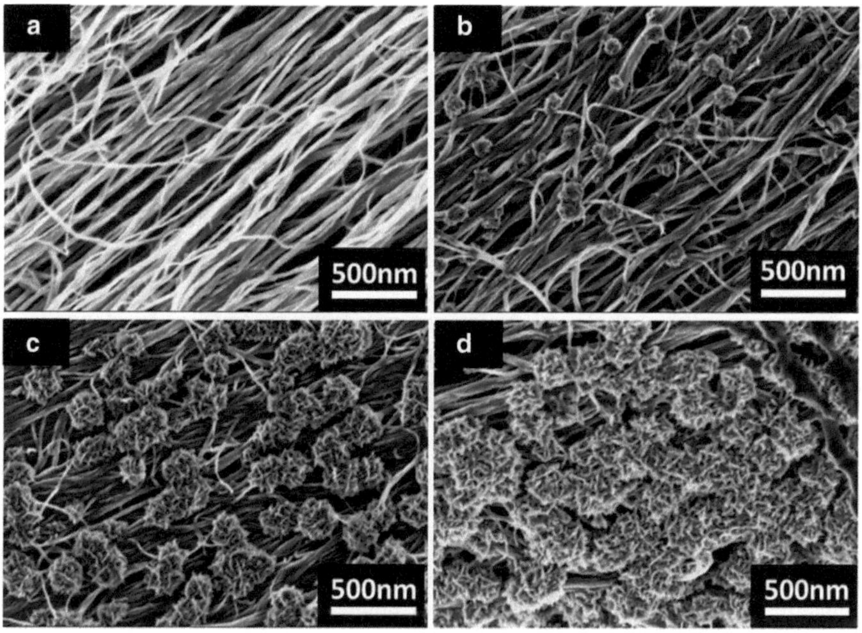

Fig. 7.8 Scanning electron microscopy (SEM) images of aligned CNT fibers before and after compositing with MnO_2 nanoparticles. **a** Bare CNT fiber. **b–d** CNT/MnO_2 hybrid fibers with MnO_2 weight contents of 0.5 %, 4.1 %, and 8.6 %, respectively (Reproduced from Ref. [26] by permission of John Wiley & Sons Ltd)

7.3.1.3 Electrochemical Performance

The voltage profiles of the composite fiber during charge and discharge processes are exhibited in Fig. 7.9, with bare CNT fiber as comparison. The low intercalation/deintercalation potential (~0.4 V) rendered the bare CNT fiber ineligible for cathode in lithium ion battery. However, for the CNT/MnO_2 composite fiber, the discharge platform was 1.5 V, and the specific capacity was 109.62 mAh cm^{-3} or 218.32 mAh g^{-1} based on the whole CNT/MnO_2 electrode at 5×10^{-4} mA. At an elevated current, the specific capacity endured a slight decrease to 94.37 mAh cm^{-3} (174.40 mAh g^{-1}) at 2×10^{-3} mA.

7.3.1.4 Summary

Due to the excellent mechanical and electronic properties of the aligned CNT/MnO_2 composite fibers, the fiber-shaped lithium ion battery displayed a good electrochemical performance. The flexible one-dimensional structure also enables promising applications pertaining to wearable devices. However, the lithium wire electrode significantly limited the practical application of such a fiber-shaped battery.

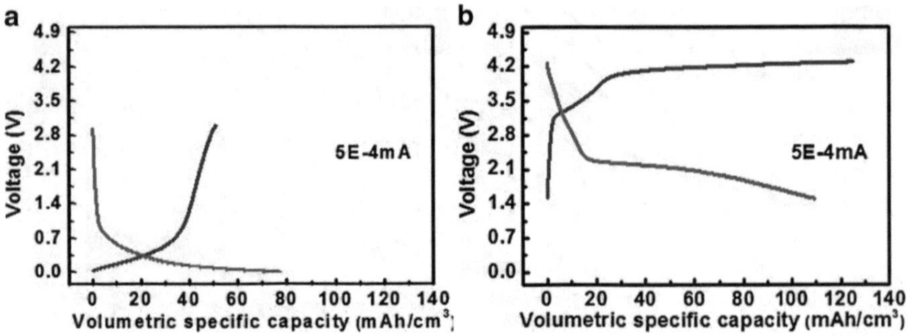

Fig. 7.9 Voltage profiles of the fiber-shaped lithium ion batteries during charge and discharge process at a current of 5×10^{-4} mA. **a** A bare CNT fiber. **b** A CNT/MnO_2 hybrid fiber with 4.1 wt % of the MnO_2. Both of the batteries used lithium wire as anode (Reproduced from Ref. [26] by permission of John Wiley & Sons Ltd)

7.3.2 Fiber-Shaped CNT/Si Anode

Silicon has been widely studied as a promising anode material with the theoretical capacity of 4,200 mAh g^{-1} and lithiation potential within 0.2–0.4 V. However, silicon anode also suffers from several tricky issues that impede its steps towards application. The salient problem is the significant volume expansion (400 %) associated with the formation of $Li_{4.4}Si$ alloy. The huge volume change on one hand generates internal stress that leads to the pulverization of anode; on the other hand, it is detrimental to the solid electrolyte interphase (SEI) film on the surface of the anode. The repeated dismantling and reconstruction of SEI film will deplete the electrolyte and thus render the battery deteriorated. Considering the safety vulnerability has made the lithium metal marginalized, the need for a competent successor motivates an intensive investigation in the silicon anode. The preliminary efforts towards fiber-shaped lithium ion battery, as we have discussed in the last section, started with the preparation of a fiber cathode. Metallic lithium wire, unwillingly though, was used as anode without any choices. As pundits touted the fiber-shaped devices as the state-of-the-art products, the flexibility is acclaimed as their highlight property distinct from their planar counterparts, which, nevertheless, is weakened with the presence of lithium metal in lithium ion batteries. Therefore, preparing an eligible fiber electrode to displace the lithium wire as anode in fiber-shaped lithium ion battery is badly needed. To this end, we proposed a fiber electrode that integrates silicon and CNT to pair with the fiber cathode in lithium ion batteries. In this section, the preparation and performances of the CNT/Si composite fiber as anode are presented [27].

7.3.2.1 Synthesis of Materials

Different from the former process, the silicon was not incorporated to the CNT fiber directly. Instead, the electron beam evaporation technology was used to

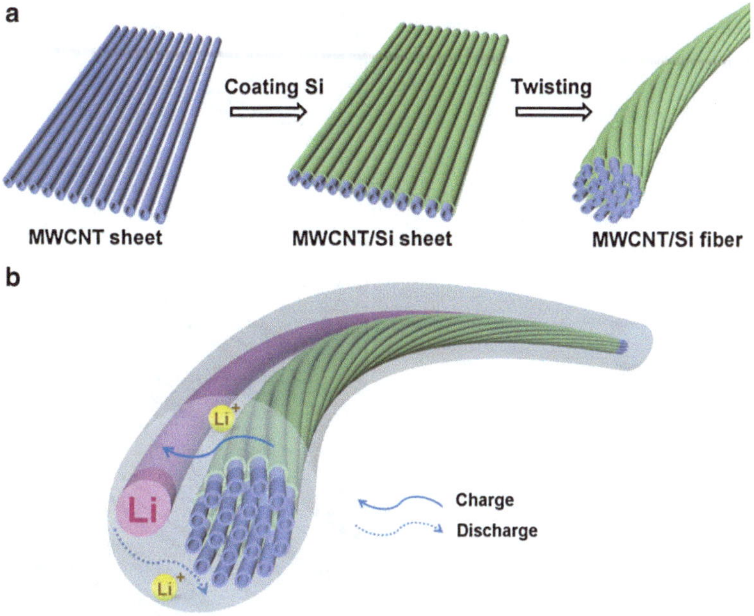

Fig. 7.10 **a** Preparation of the CNT/Si hybrid fiber. **b** Schematic illustration to the fiber-shaped lithium ion battery based on the CNT/Si hybrid fiber (Reproduced from Ref. [27] by permission of John Wiley & Sons Ltd)

vaporize the silicon which then was deposited onto the CNT sheet, an aligned two-dimensional film that had been drawn from the spinnable CNT array and paved on a substrate. The amount of the silicon deposited was varied by the sputtering time. Then the composite films were twisted into CNT/Si composite fiber at a rotating speed of 180 rounds per minute. The preparation process is illustrated in Fig. 7.10a. To examine its electrochemical performance, the CNT/Si composite fiber was first paired with lithium wire in a battery with the conventional LB303 electrolyte (the components of LB303 was mentioned in Sect. 7.3.1) (Fig. 7.10b).

7.3.2.2 Structure

An obvious change after depositing silicon is the diameter of the fiber, which nearly doubled from 30 to 60 μm (Fig. 7.11). The CNTs in the composite fiber remained orientated as expected. The silicon was uniformly coated on the surface of the individual CNT with no aggregation detected, exhibiting a typical core-sheath structure. The content of Si in the composite fiber was 38.1 %, which is calculated from the depositing parameter and backed up by the results from energy dispersive X-ray spectroscopy (EDX). The thickness of the silicon layer was ~20 nm estimated from TEM images. A broad peak at 470 cm^{-1} in the Raman spectrum indicated the silicon deposited mainly appeared in amorphous structure.

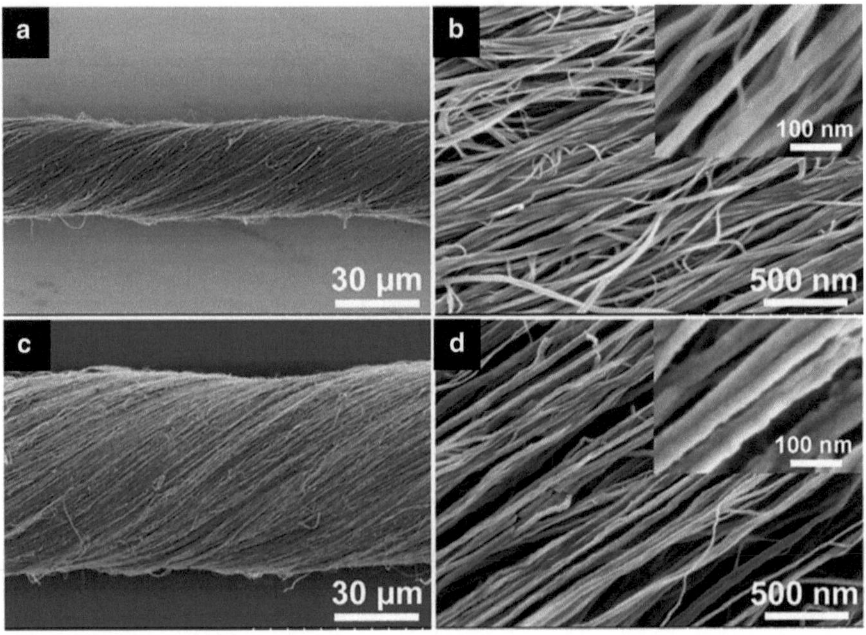

Fig. 7.11 SEM images of bare CNT and CNT/Si hybrid fiber. **a** and **b** SEM images of a bare CNT fiber at low and high magnifications, respectively. **c** and **d** SEM images of a CNT/Si hybrid fiber with 38.1 wt% of Si at low and high magnifications, respectively (Reproduced from Ref. [27] by permission of John Wiley & Sons Ltd)

The composite fibers exhibited a decent flexibility that enabled the fiber to endure the repeated deformation without detectable damage in structure. In addition, the core-sheath-structured hybrid fiber benefited from the synergetic effect of CNTs and silicon that can effectively exploit lithium storage capability of silicon. Meanwhile, the space in the hybrid fiber can accommodate the volume expansion of the silicon and alleviate its negative effect to the battery longevity, as discussed next.

7.3.2.3 Electrochemical Performance

The CNT/Si hybrid fiber was paired with lithium wire and encased in a testing mold. The electrochemical performance was tested within the voltage window of 0.02–1.20 V. The CNT/Si hybrid fiber delivered a voltage plateau around 0.4 V. Compared with the bare CNT fiber, the hybrid fiber delivered a significantly higher capacity, which increased in line with the silicon content (Fig. 7.12). It should be noted that the capacity is specified to the amount of the active materials. However, the capacity retained by 32 % after 100 cycles of draining and recharging, which was attributed to the volume change of silicon. In this work, the researcher also compared the electrochemical performance of silicon deposited on a copper foil.

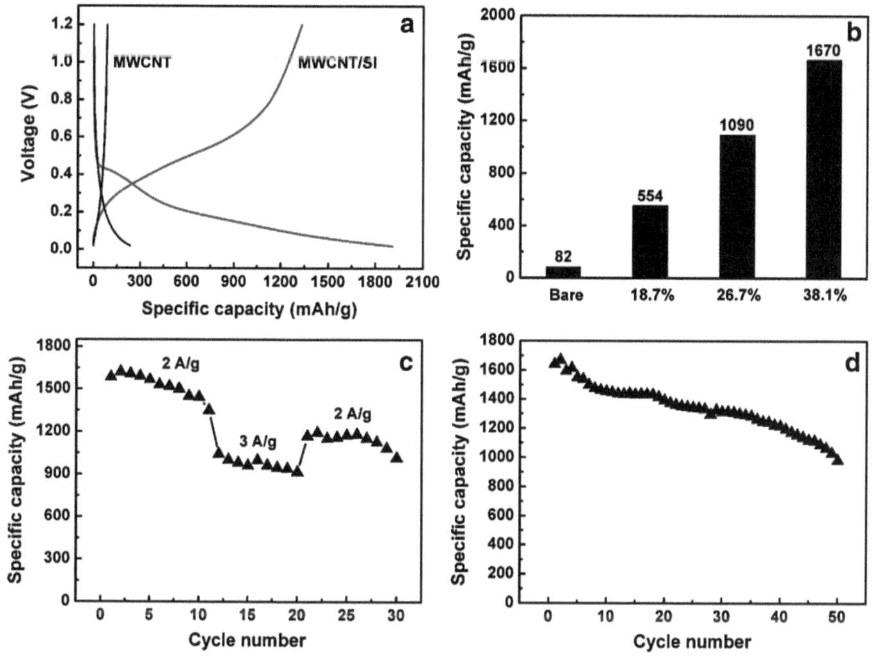

Fig. 7.12 Electrochemical performances of CNT/Si hybrid fiber. **a** Voltage profiles of a bare CNT and CNT/Si hybrid fiber. **b** Specific capacities of hybrid fibers with different silicon contents. **c** and **d** Rate and cyclic performance of the CNT/Si hybrid fiber, respectively. All the tests were conducted at 1 A g^{-1} and the CNT/Si hybrid fiber had a silicon content of 38.1 wt% (Reproduced from Ref. [27] by permission of John Wiley & Sons Ltd)

The bare silicon exhibited a high capacity of 3,537 mAh g^{-1} in the first cycle, which, nevertheless, decayed seriously and only maintained 46.6 % after 30 cycles while the CNT/Si composite fibers exhibited a capacity retention of 58 % over 50 cycles.

The CNT/Si hybrid fiber exhibited flexibility that enables itself to sustain repeated bending for 100 times without sacrificing its electrochemical performance. In fact, the capacity retained by 94 % after enduring deformations which survived more than 80 % after 20 cycles.

7.3.2.4 Summary

In summary, the CNT/Si hybrid fiber exhibited flexibility and acceptable electrochemical performances benefited from the combination effect between the CNT and silicon. The CNT/Si hybrid fiber proved its competence to displace lithium wire as a fiber anode in fiber-shaped lithium ion batteries.

7.3.3 *Fiber-Shaped $LiMn_2O_4$–Si Battery*

The success of CNT/Si hybrid fiber as electrode leapt the hurdle of the safety vulnerability of metallic lithium. Therefore, pairing the fiber cathode with the nonmetallic lithium anode to fabricate a full battery in fiber shape is naturally on the to-do list.

7.3.3.1 Synthesis of Materials

Since $LiMn_2O_4$ (LMO) was widely researched as the cathode of lithium ion battery due to its high working voltage (~4 V) and structural stability (volume change less than 10 %), in this full battery, CNT/LMO hybrid fiber was used as cathode [28]. The CNT/Si hybrid fiber was used as an anode to replace metallic lithium. Both of the two electrodes were oriented from CNT sheet as displayed in Fig. 7.13a, b. $LiMn_2O_4$ particles were directly impregnated into the aligned CNT sheet through a solution-based process, followed by twisting into a hybrid fiber. In light of the dramatic volume change of Si, a hybrid layered structure was developed to better the performance of CNT/Si anode. Si was first deposited onto the CNT sheet by electron beam evaporation, which was then sandwiched between two bare CNT sheets and further scrolled into a hybrid fiber in a hybrid layered structure. As showed in Fig. 7.15a, the fiber-shaped lithium ion battery was fabricated by winding two aligned CNT hybrid fibers onto a cotton fiber, followed by coating gel electrolyte.

7.3.3.2 Structure

Si was uniformly coated on the surfaces of aligned CNTs to form a coaxial structure. The CNTs and coaxial CNT/Si nanotubes were oriented in the CNT/Si composite fibers (Fig. 7.13c, d). The diameter of the CNT/Si composite fiber was around 100 μm. The Si content of 62 % was given by thermogravimetric analysis (TGA). As shown in Fig. 7.13e, f, after deposition of $LiMn_2O_4$ particles, the CNTs also remained aligned and the diameter of the composite fibers was about 100 μm. The LMO particles had an average size of 400 nm and accounted for 75 % of the electrode weight.

7.3.3.3 Electrochemical Performance

The electrochemical performances of the two hybrid fiber electrodes were tested in the half cells against lithium. Typical cyclic voltammograms (CV) of CNT/Si composite fiber recorded at 0.1 $mV\ s^{-1}$ were displayed in Fig. 7.14a. The formation of solid electrolyte interface (SEI) was reflected in the first lithiation curve where a

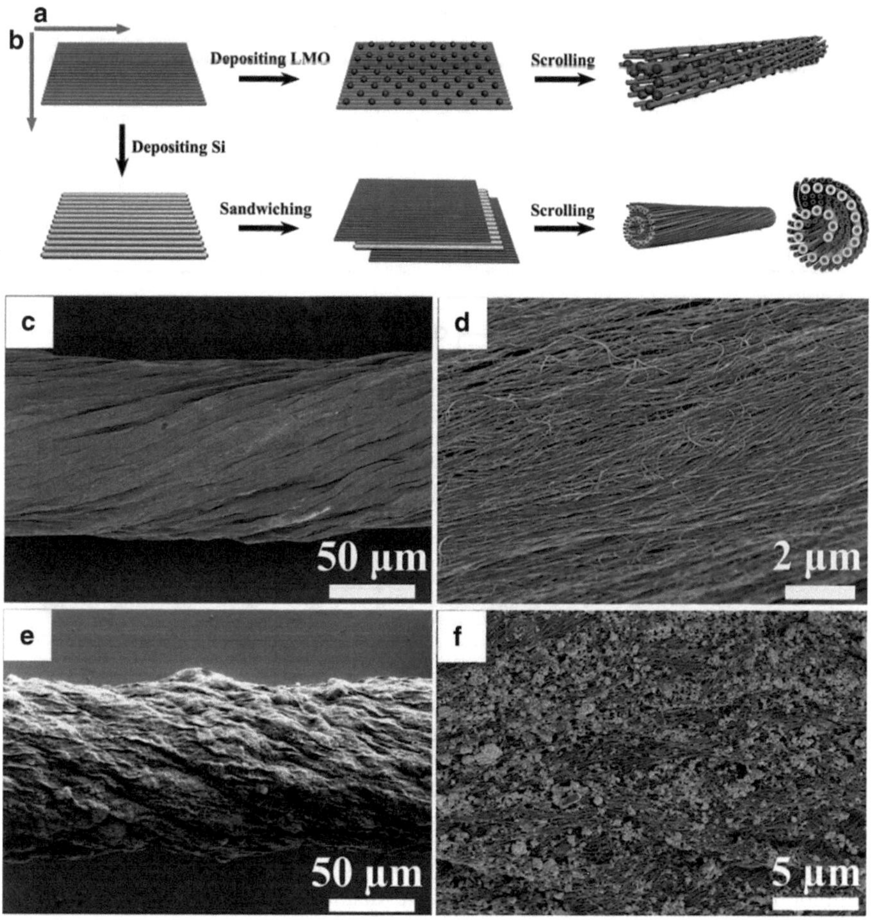

Fig. 7.13 Preparation and morphology characterization of the CNT hybrid fiber. **a** CNT/LMO fiber. **b** CNT/Si fiber. **c** and **d** SEM images of CNT/Si fiber at low and high magnifications, respectively. **e** and **f** SEM images of CNT/LMO fiber at low and high magnifications, respectively (Reprinted with the permission from Ref. [28]. Copyright 2014 American Chemical Society)

current peak appeared between 0.5 and 1.0 V. The phase transition in amorphous Li_xSi and formation of crystalline $Li_{15}Si_4$ corresponded to the peaks at 0.17 and 0.04 V, while the broad anodic peak at 0.6 V referred to a phase transition from amorphous Li_xSi to amorphous silicon. The initial delithiation capacity (charge capacity) was 2,240 mAh g^{-1} at 0.4 C, which retained 88 % of its capacity after 100 cycles. The CNT skeleton provided effective conducting pathways, making the battery adaptable for large current. As a demonstration, at a high current rate of 2 C (8,400 mA g^{-1}), the CNT/Si hybrid fiber delivered a high delithiation capacity of 1,523 mAh g^{-1} and more than 85 % of the capacity survived after running for 400 cycles (Fig. 7.14b). The much improved stability was benefited from the hybrid

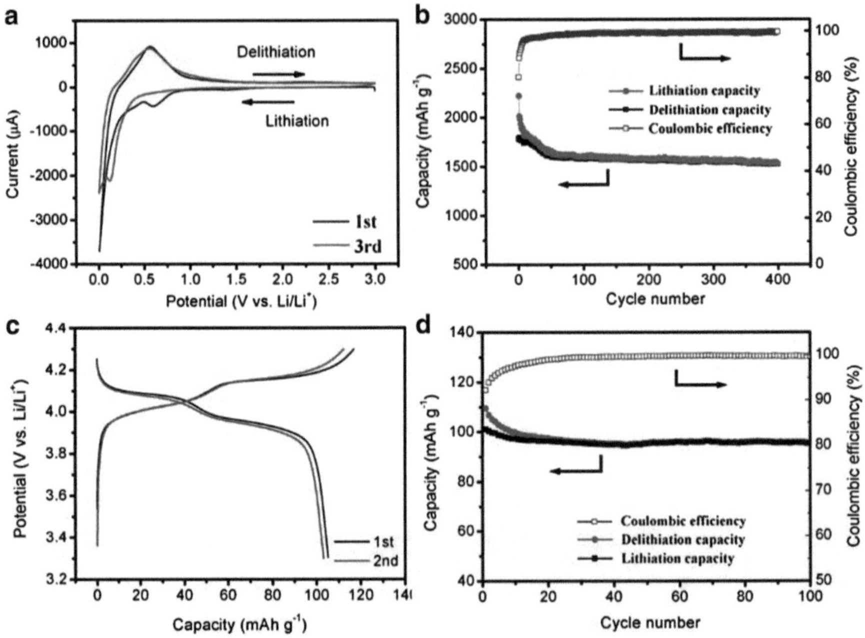

Fig. 7.14 Electrochemical performances of CNT/Si and CNT/LMO hybrid fibers. **a** and **b** Cyclic voltammogram and long-life performance (2 C) of CNT/Si hybrid fiber, respectively. **c** and **d** Voltage profiles (0.5 C) and long-life performance (1 C) of CNT/LMO hybrid fiber, respectively (Reprinted with the permission from Ref. [28]. Copyright 2014 American Chemical Society)

layered structure where CNT/Si hybrid sheet was sandwiched between bare CNT sheets. The CNT sheets were conducive to buffer the volume expansion and immobilize the silicon layer. Figure 7.14c displays the voltage profiles of the CNT/LMO hybrid fiber at 0.5 C where two distinct plateaus were observed in the charge and discharge curves, indicating a reversible two-step transition spanning from λ-MnO_2, $Li_{0.5}Mn_2O_4$ to $LiMn_2O_4$. The initial discharge capacity was 101 mAh g^{-1} with a Coulombic efficiency of 92 %, which retained 94 % of its capacity after 100 cycles (Fig. 7.14d).

The full battery does not need a pre-lithiation process since the lithium atoms in $LiMn_2O_4$ are available in the electrochemical process. The fresh battery was in a drained state and should be charged first. During charge process, the lithium atoms were removed from the cathode, $LiMn_2O_4$, and the lithium ions compounded with Si at the anode. When the full battery was discharged, it reversed. The capacity matchability of the anode and cathode is a critical issue that dominates the performance of a full battery. Ideally, the lithium stored in anode and cathode should be equal to ensure that lithium removed from cathode can fully intercalate into the anode. However, in practical circumstances, the charge and discharge process of a lithium ion battery always involves the reaction with electrolyte, which consumes lithium ions and forms solid electrolyte interface (SEI). As clarified by Eq. 7.3 and

Fig. 7.3, an excess capacity in anode is conducive to exploit the total capacity of a battery. However, if the anode has a much larger capacity, it consumes more lithium ions form SEI which brings about large irreversible capacity. Therefore, the anode and cathode should be paired in a delicate way to keep the anode capacity slightly exceeded. Note that the capacity here refers to the actual capacity relating to the lithium ions stored in the electrode. The fiber-shaped full battery had a length density of 27 mg cm^{-1} including 0.18 mg cm^{-1} of Si and 2.1 mg cm^{-1} LMO. The capacity ratio between anode and cathode was settled as 1.5:1, which was controlled based on the length density and stable capacity after running for several cycles.

The fiber-shaped lithium ion battery was performed in a gel electrolyte with an output voltage of 3.4 V, in line with the voltage difference between LMO (4.0 vs. Li/Li^+) and Si (0.6 vs. Li/Li^+). The initial specific capacity is 106.5 mAh g^{-1} (specified to the CNT/LMO cathode), which retained 87 % of its capacity after 100 cycles (Fig. 7.15b, c). The fiber-shaped full battery can stably run for hundreds of draining and recharging processes, indicating the capacities of two electrodes

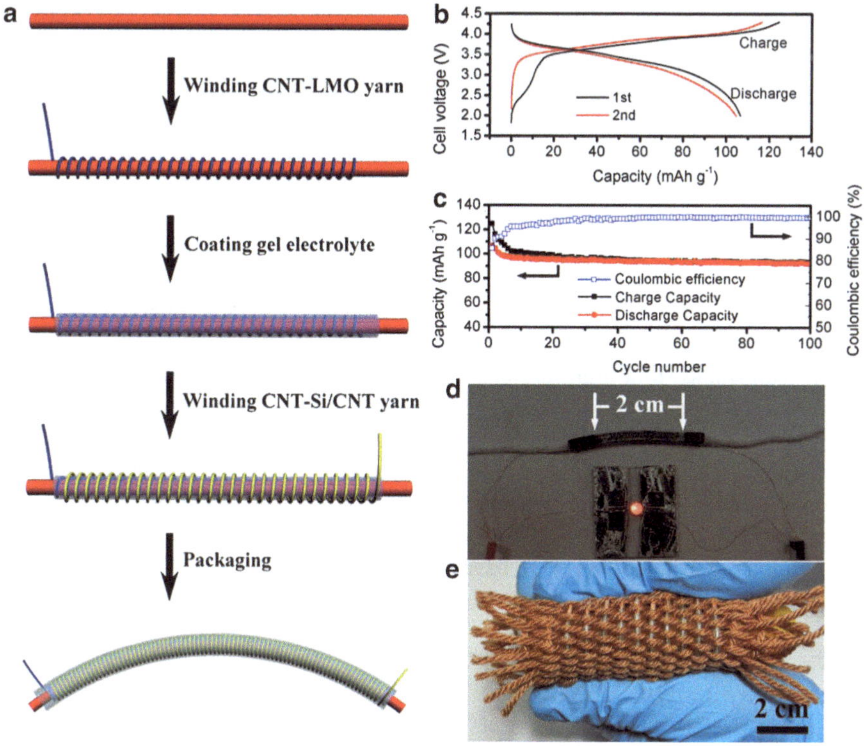

Fig. 7.15 **a** Fabrication of the fiber-shaped lithium ion battery based on a cotton wire. **b** Voltage profiles at 1 C. **c** Long-life performance at 1 C. **d** Photograph of a fiber-shaped lithium ion battery powering an LED. **e** Fiber-shaped lithium ion batteries being woven into a cloth (Reprinted with the permission from Ref. [28]. Copyright 2011 American Chemical Society)

were well matched. As an exemplification to application, a 2 cm of fiber-shaped battery powering an LED is displayed in Fig. 7.15d. Furthermore, the fiber battery can be woven into a cloth for wearable devices (Fig. 7.15e).

7.3.3.4 Conclusion

The fiber-shaped lithium ion battery based on CNT/LMO and CNT/Si hybrid fibers exhibited remarkable electrochemical performances. It was also metal free, light weight, and flexible and can be woven into fabric. However, the full battery did not take advantages of the high capacity of Si electrode, and the capacity of the full battery was only 106.5 mAh g^{-1}. In addition, the fabrication process was complicated involving a capacity match between the anode and cathode. Therefore, further improvement is underway.

7.3.4 *Fiber-Shaped $LiMn_2O_4$–$Li_4Ti_5O_{12}$ Battery*

Replacing metallic lithium with alloy anode is marked as an advancement since it, for the first time, makes fiber-shaped lithium ion batteries free from the safety vulnerability and more conformable to the concept of wearability. The silicon anode has a high theoretical capacity, which is conducive to squeeze the capability of cathode and increase the total capacity of battery. Nevertheless, its fragile performance and laborious preparation weaken the reliability of battery. On this account, we shift our focus to a commercialized anode material: $Li_4Ti_5O_{12}$.

$Li_4Ti_5O_{12}$ (LTO) is regarded as a zero-strain material for lithium intercalation which suggests that no volume change occurs during the lithiation and delithiation process. The ions migration will not be blocked by the transformation of materials, which contributes to a high rate performance and high cyclic stability. $Li_4Ti_5O_{12}$ has a working potential at 1.5 V (vs. Li/Li^+) which is above the electrochemical decomposition potential for most of the common electrolytes, suppressing the formation of SEI and promising a better longevity. $Li_4Ti_5O_{12}$ has been commercialized to couple with $LiMn_2O_4$ (LMO), which has a working potential of 4 V against lithium, with a voltage plateau around 2.5 V. The presented fiber-shaped lithium ion battery is constructed based on LTO/LMO battery, with CNT/LTO and CNT/LMO fiber electrodes [29].

7.3.4.1 Synthesis of Materials

$LiMn_2O_4$ nanoparticles were synthesized by the hydrothermal method, while $Li_4Ti_5O_{12}$ nanoparticles were obtained by a solid-state method followed by ball milling. The X-ray diffraction (XRD) pattern confirmed the as-prepared $LiMn_2O_4$ and $Li_4Ti_5O_{12}$ nanoparticles have a spinel structure, and scanning electron microscopic (SEM) images showed both lithium metal oxides had hundreds of nanometers in diameter. The key to fiber-shaped battery lies in the fiber electrode. Leveraging the superiority of CNTs, the active materials were incorporated in CNT fibers through a co-spinning process. Typically, $LiMn_2O_4$ and $Li_4Ti_5O_{12}$ nanoparticles were dispersed in *N,N*-dimethylformamide (DMF) to form uniform suspensions. Then, two CNT sheets drawn out of the CNT array were stacked and subsequently immersed into the as-prepared nanoparticle suspensions. The active materials will be embraced in the CNT fiber during the co-spinning process. As we have discussed in Chap. 2, the aligned CNTs sheets are prone to shrink when encountering many solvents like water and alcohol. DMF has been proved a reliable surrogate which can disperse nanoparticles and where CNTs can stretch like ribbon. Particularly, for the $LiMn_2O_4$ suspension, a certain amount of CNTs was added to help anchor the $LiMn_2O_4$ nanoparticles onto the hybrid fibers. CNT/LTO hybrid fibers were coated with a thin layer of graphene oxide (GO) through dip coating to improve structural stability.

7.3.4.2 Structure

Figure 7.16 shows the SEM images of CNT/LTO and CNT/LMO composite fibers. It is revealed that both CNT/LMO and CNT/LTO hybrid fibers have

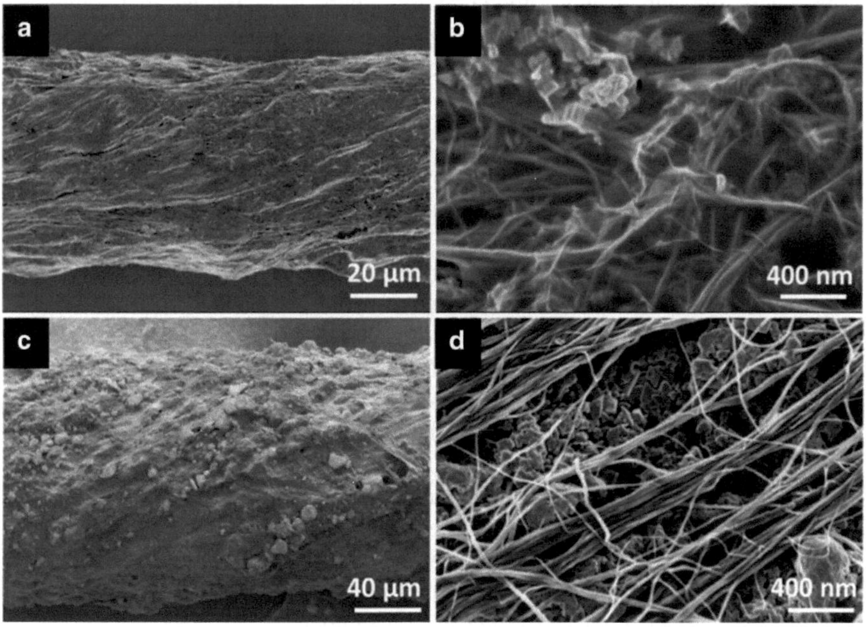

Fig. 7.16 SEM images of CNT/LTO hybrid fiber (**a** and **b**) and CNT/LMO hybrid fiber (**c** and **d**) (Reproduced from Ref. [29] by permission of John Wiley & Sons Ltd)

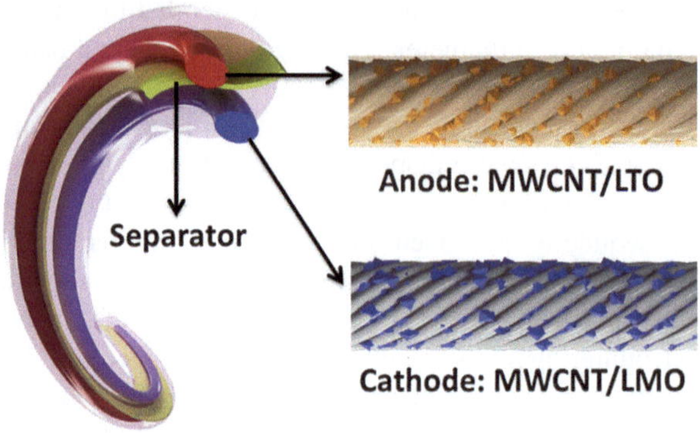

Fig. 7.17 Schematic illustration to the fiber-shaped lithium ion battery (Reproduced from Ref. [29] by permission of John Wiley & Sons Ltd)

uniform diameters of 130 and 70 μm, respectively. These nanoparticles are evenly incorporated within CNT fibers. The added CNTs and GO layer can be respectively found on the surface of two hybrid fibers. They can form a network structure to prevent the nanoparticles divorcing from the framework during charge and discharge process, which helps to improve the cyclic stability. The composite fibers have linear densities of 2 and 10 mg m^{-1} for the CNT/LTO and CNT/LMO fibers, respectively. The thermogravimetric analysis (TGA) showed the contents of $LiMn_2O_4$ and $Li_4Ti_5O_{12}$ within the fibers were 90 % and 78 %, respectively. The hybrid fibers were flexible and strong to sustain various deformations without structural damage. The fiber-shaped battery was assembled in a heat-shrinkable tube where the anode and cathode hybrid fibers were placed in parallel and separated by a polyvinylidene fluoride (PVDF) membrane. Finally, the conventional liquid electrolyte (LB303) was injected into the tube in an argon-filled glove box (Fig. 7.17).

7.3.4.3 Electrochemical Performance

In the beginning, the electrochemical performance of the anode and cathode fibers was examined in half cells with the lithium metal as counter electrodes. Figure 7.18a, c displays the charge and discharge profiles for both hybrid fibers. The specific capacities for CNT/LMO and CNT/LTO hybrid fibers were respectively calculated as 60 and 150 mAh g^{-1}, lower than their theoretical values. As expected, the discharge plateau for the CNT/LTO composite fiber was 1.5 V, while the charge plateau for the CNT/LMO was 4.0 V. After running for 200 cycles, both of the half batteries retained their capacities above 80 %, much higher than the silicon anode-based fiber-shaped batteries (Fig. 7.18b, d). The effects of GO and CNTs were revealed by controlled experiments. For the CNT/LTO hybrid fiber, only 60 % of capacity left in absence of GO. The added CNTs can effectively improve the rate performance of the CNT/LMO fiber.

The CNT/LTO and CNT/LMO fibers both exhibited stable electrochemical behaviors. Their combination, the full cell, was well performed with respect to

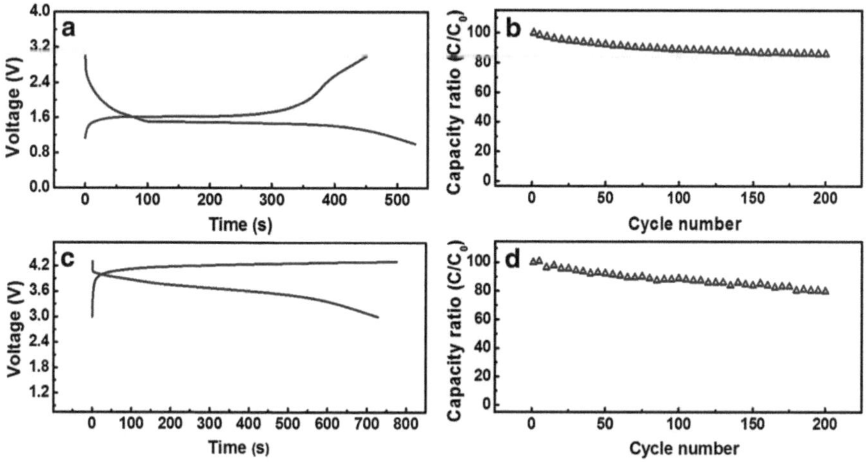

Fig. 7.18 Voltage profiles and long-life capacity retention of CNT/LTO (**a** and **b**) and CNT/LMO (**c** and **d**) hybrid fibers in half cells. The voltage profiles were obtained at 0.02 mA and long-life tests were run at 0.05 mA. The lengths of CNT/LTO and CNT/LMO hybrid fibers were 1 cm in the half battery (Reproduced from Ref. [29] by permission of John Wiley & Sons Ltd)

the specific capacities and voltage plateaus. The fiber-shaped LTO/LMO battery had a specific capacity of 138 mAh g^{-1} at 0.01 mA (specified to the weight of the CNT/LTO fiber). It also delivered a volumetric energy density of 17.7 Wh L^{-1} and power density of 560 W L^{-1} based on the volume of anode and cathode. Besides, the capacity retained by 85 % after 100 cycles at 0.05 mA, with Coulombic efficiency over 80 %.

7.3.4.4 Flexibility and Stretchability

Benefited from the flexible components, the fiber-shaped battery exhibited good flexibility and can be bent into various shapes without fracture in structure and degradation in performances (Fig. 7.19a, b). Moreover, the battery can be scaled up to 200 cm as shown in Fig. 7.19d. As a demonstration to practical use, a 10 cm charged fiber-shaped battery can power nine red light emitting diodes (LEDs) for at least one minute (Fig. 7.19c). As we have proposed, the fiber-shaped batteries are developed in orientation towards wearable devices. In this regards, an energy fabric was woven from fiber-shaped lithium ion batteries which was then integrated into a cloth (Fig. 7.19e–g).

In order to deal with the stretch deformation during use, the fiber-shaped lithium ion battery should be endowed with stretchability. However, this project encounters many challenges since the CNT-based electrode and liquid electrolyte are inadaptable to stretch deformation. An ingenious approach based on structure design is presented here. The anode and cathode fibers were fixed in parallel and

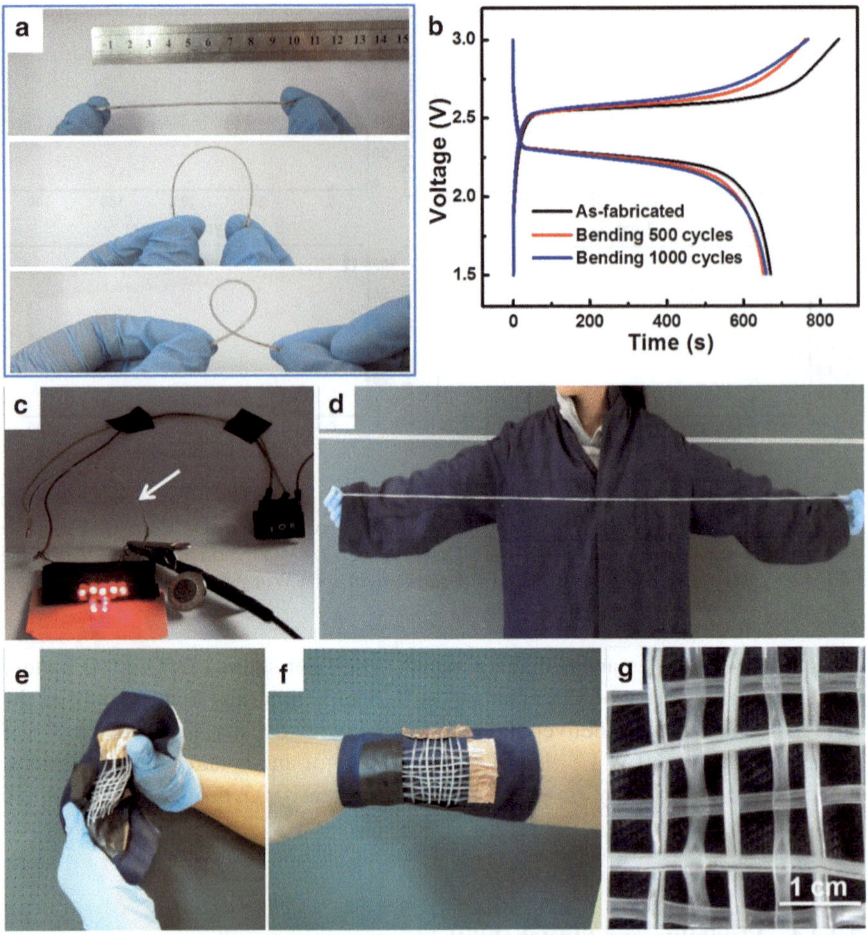

Fig. 7.19 Flexibility of fiber-shaped lithium ion batteries. **a** Photographs of a fiber-shaped battery being deformed into different shapes. **b** Voltage profiles of the battery before and after bending for 500 and 1,000 cycles at 0.05 mA. **c** Nine LEDs being powered by a fiber-shaped battery with a length of 10 cm. **d** A long fiber-shaped lithium ion battery with a length of 200 cm. **e–g** Fiber-shaped batteries woven into energy fabrics (Reproduced from Ref. [29] by permission of John Wiley & Sons Ltd)

spirally wound in helix onto an elastic substrate, followed by coating of a thin layer of gel electrolyte and sealing into a heat-shrinkable tube. As illustrated in Fig. 7.20, the stretchability of the battery was derived from the elastic substrate. The spirally wound electrodes bypassed the dilemma that CNT/LTO and CNT/LMO hybrid fibers were not elastic. This strategy is in line with the methods used in stretchable dye-sensitized solar cells (see Chap. 3). When the battery is stretched, the interspace between the anode and cathode expands but the battery retains its configuration. The gel electrolyte plays a critical role during stretching since the

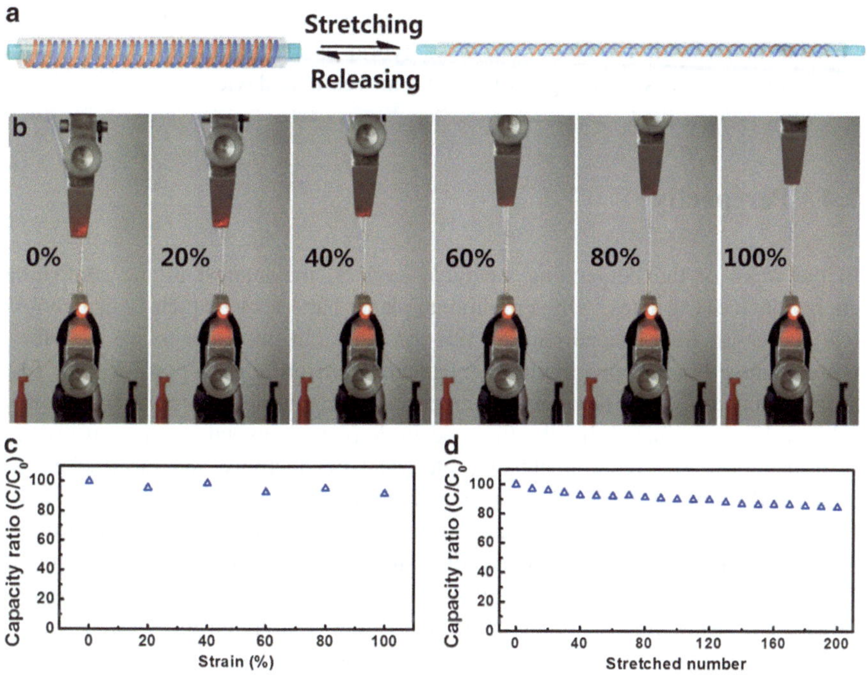

Fig. 7.20 Stretchability of fiber-shaped lithium ion batteries. **a** Schematic illustration to the battery configuration during stretching. **b** Photographs of a stretched battery powering an LED. **c** Capacity evolution with stretch strain. **d** Capacity evolution with stretching number (stretching strain: 100 %). C_0 and C correspond to the specific capacities before and after stretching (Reproduced from Ref. [29] by permission of John Wiley & Sons Ltd)

polymer gel is conformable to deformation and prevents the electrodes from slipping off. The electrochemical performances of the battery have been investigated under different stretching deformations. The specific capacity retained above 90 % when stretched by 100 %. After repeating for 200 cycles, more than 80 % of capacity remained. This configuration is also adaptable to large elongation provided that elastic substrate is replaced by a rubbery fiber, polydimethylsiloxane (PVDF), for example. As a result, the battery enabled a high stretching strain up to 600 % [30].

7.3.4.5 Summary

Through the assembly of two hybrid fibers, a fiber-shaped lithium ion battery was fabricated. Without the metallic lithium as anode, this full battery was immune from the safety concerns that have embarrassed its predecessors. It has good cyclic and rate performance, and its volumetric energy densities and power densities were higher than the lithium thin-film batteries. In the perspective of mechanical properties, it exhibits decent flexibility and can be woven into a fabric. Moreover, through the strategic configuration, these hybrid fibers can be used to fabricate a stretchable battery, showing great potential in wearable devices.

7.4 Perspective

In the wake of the burgeoning wearable devices, the interest to the compatible energy devices escalates. As the traditional planar batteries lose their lure, a competent successor, fiber-shaped battery, fills the blank. In this chapter, we describe a representative fiber-shaped lithium ion battery. Transforming the planarity into fiber shape is not easy, even formidable when encountering liquid electrolyte and metallic lithium. It is recognized that the fiber electrode plays a pivotal part in constructing fiber batteries. The requirements for electrical conductivity and electrochemical compatibility make CNT fiber, a new recruit in fiber materials, edge ahead. Rudimentary prototypes notwithstanding several fiber-shaped lithium ion batteries along with their functionalized derivative, stretchable batteries have been materialized leveraging the CNT fibers. Ingenious strategies including replacing metallic lithium and using gel electrolyte contribute to the contraption that enables energy storage and delivery in a light, flexible, and miniature battery.

However, roads towards market are still rugged. A foreseeable challenge is, in fact, the inferior performances. Skipping the results highlighted in papers, a practical issue is that the small contents of active materials in fiber electrode cannot deliver a large energy storage capacity. In spite of a reasonable specific capacity or energy density on par with planar counterparts, the fiber-shaped lithium ion battery can only afford a blink of bulb since the energy stored is depleted quickly. However, the challenge meanwhile paints an encouraging picture for improvement. In light of the staggering increase in the capacity of the conventional lithium ion batteries, the performance of the fiber-shaped battery, in consequence, has its ceiling. Therefore, an upgrade in capacity and energy should rely on the revolution in active materials with higher theoretical capabilities. Although silicon has been introduced in fiber-shaped batteries as anode, the dwarfed cathode capacity can hardly exploit the superiority of the silicon. Sulfur, with lower cost and much higher theoretical capacity and energy density, is probably an eligible candidate. Another problem lies in the ionic conductor, the electrolyte. A fiber-shaped lithium ion battery with liquid electrolyte can hardly conform the concept of wearable device and it is farfetched to claim to be flexible. The liquid electrolyte carries a risk of leakage not least under deformation. Moreover, it makes the battery less adaptable to the environment since the temperature window for operation is ring-fenced by the solvent in electrolyte. On this account, gel electrolyte is more suitable for fiber-shaped lithium ion batteries. But the sluggish ion migration in polymer gel causes large polarization and reduces the rate performance.

The problems that fiber-shaped lithium ion battery is suffering have plagued the conventional lithium ion batteries for years. The breakthrough in conventional lithium ion batteries also benefits their fiber-shaped counterparts. Individually, for fiber-shaped lithium ion battery, efforts can be devoted to elongating the battery while retaining its performance, which requires the concerted progress in electrode preparation and device fabrication. Besides, the fiber-shaped battery endures more rapid degradation, partially arising from the device fabrication process such as sealing. On the other hand, the influence of one-dimensional configuration of the battery also needs elucidation. The importance of fiber-shaped lithium ion battery is widely appreciated because of its unique structure and potential application. In the near future, the design and exploitation of fiber-shaped lithium ion battery should associate with the consideration in user experience and practical requirements.

References

1. Abu-Lebdeh Y (2013) Introduction. In: Davidson I, Abu-Lebdeh Y (eds) Nanotechnology for lithium-ion batteries. Springer, New York, pp 2–11
2. Tarascon JM, Armand M (2001) Issues and challenges facing rechargeable lithium batteries. Nature 414(6861):359–367
3. Imanishi N, Ohashi S, Ichikawa T, Takeda Y, Yamamoto O, Kanno R (1992) Carbon lithium anodes for lithium secondary batteries. J Power Sources 39(2):185–191
4. Yoshio M, Noguchi H (2009) A review of positive electrode materials for lithium-ion batteries. In: Yoshio M, Brodd RJ, Kozawa A (eds) Lithium-Ion batteries: science and technologies. Springer, New York, p 11
5. Ferg E, Gummow R, De Kock A, Thackeray M (1994) Spinel anodes for lithium-ion batteries. J Electrochem Soc 141(11):L147–L150
6. Matsumura Y, Wang S, Mondori J (1995) Interactions between disordered carbon and lithium in lithium ion rechargeable batteries. Carbon 33(10):1457–1462
7. Babu G, Kalaiselvi N, Bhuvaneswari D (2014) Synthesis and surface modification of LiCo1/3Ni1/3Mn1/3O2 for lithium battery applications. J Electron Mater 43(4):1062–1070
8. Singh G, Thomas R, Kumar A, Katiyar R (2012) Electrochemical behavior of Cr-doped composite Li_2MnO_3-$LiMn_{0.5}Ni_{0.5}O_2$ cathode materials. J Electrochem Soc 159(4): A410–A420
9. Whittingham MS (2004) Lithium batteries and cathode materials. Chem Rev 104(10):4271–4302
10. Li ZH, Wang LQ, Li KY, Xue DF (2013) $LiMn_2O_4$ rods as cathode materials with high rate capability and good cycling performance in aqueous electrolyte. J Alloys Compd 580:592–597
11. Pan F, W-l W, Li H, Xin X, Chang Q, Yan W, Chen D (2011) Investigation on a core–shell nano-structural $LiFePO_4$/C and its interfacial C-O interaction. Electrochim Acta 56(20):6940–6944
12. Jin Y, Yang CP, Rui XH, Cheng T, Chen CH (2011) V_2O_3 modified $LiFePO_4$/C composite with improved electrochemical performance. J Power Sources 196(13):5623–5630
13. Zhao Y, Peng LL, Liu BR, Yu GH (2014) Single-crystalline $LiFePO_4$ nanosheets for high-rate Li-ion batteries. Nano Lett 14(5):2849–2853
14. Dey A (1971) Electrochemical alloying of lithium in organic electrolytes. J Electrochem Soc 118(10):1547–1549
15. Etacheri V, Marom R, Elazari R, Salitra G, Aurbach D (2011) Challenges in the development of advanced Li-ion batteries: a review. Energy Environ Sci 4(9):3243

16. Szczech JR, Jin S (2011) Nanostructured silicon for high capacity lithium battery anodes. Energy Environ Sci 4(1):56
17. Zhang W-J (2011) A review of the electrochemical performance of alloy anodes for lithium-ion batteries. J Power Sources 196(1):13–24
18. Xu K (2004) Nonaqueous liquid electrolytes for lithium-based rechargeable batteries. Chem Rev 104(10):4303–4417
19. Hu LB, Wu H, La Mantia F, Yang YA, Cui Y (2010) Thin, flexible secondary Li-ion paper batteries. ACS Nano 4(10):5843–5848
20. Li N, Chen ZP, Ren WC, Li F, Cheng HM (2012) Flexible graphene-based lithium ion batteries with ultrafast charge and discharge rates. Proc Natl Acad Sci U S A 109(43):17360–17365
21. Liu B, Zhang J, Wang XF, Chen G, Chen D, Zhou CW, Shen GZ (2012) Hierarchical three-dimensional $ZnCo_2O_4$ nanowire arrays/carbon cloth anodes for a novel class of high-performance flexible lithium-ion batteries. Nano Lett 12(6):3005–3011
22. Koo M, Park KI, Lee SH, Suh M, Jeon DY, Choi JW, Kang K, Lee KJ (2012) Bendable inorganic thin-film battery for fully flexible electronic systems. Nano Lett 12(9):4810–4816
23. Xu S, Zhang YH, Cho J, Lee J, Huang X, Jia L, Fan JA, Su YW, Su J, Zhang HG, Cheng HY, Lu BW, Yu CJ, Chuang C, Kim TI, Song T, Shigeta K, Kang S, Dagdeviren C, Petrov I, Braun PV, Huang YG, Paik U, Rogers JA (2013) Stretchable batteries with self-similar serpentine interconnects and integrated wireless recharging systems. Nat Commun 4:1543
24. Song ZM, Ma T, Tang R, Cheng Q, Wang X, Krishnaraju D, Panat R, Chan CK, Yu HY, Jiang HQ (2014) Origami lithium-ion batteries. Nat Commun 5:3140
25. Kwon YH, Woo SW, Jung HR, Yu HK, Kim K, Oh BH, Ahn S, Lee SY, Song SW, Cho J, Shin HC, Kim JY (2012) Cable-type flexible lithium ion battery based on hollow multi-helix electrodes. Adv Mater 24(38):5192–5197
26. Ren J, Li L, Chen C, Chen XL, Cai ZB, Qiu LB, Wang YG, Zhu XR, Peng HS (2013) Twisting carbon nanotube fibers for both wire-shaped micro-supercapacitor and micro-battery. Adv Mater 25(8):1155–1159
27. Lin HJ, Weng W, Ren J, Qiu LB, Zhang ZT, Chen PN, Chen XL, Deng J, Wang YG, Peng HS (2014) Twisted aligned carbon nanotube/silicon composite fiber anode for flexible wire-shaped lithium-Ion battery. Adv Mater 26(8):1217–1222
28. Weng W, Sun Q, Zhang Y, Lin HJ, Ren J, Lu X, Wang M, Peng HS (2014) Winding aligned carbon nanotube composite yarns into coaxial fiber full batteries with high performances. Nano Lett 14(6):3432–3438
29. Ren J, Zhang Y, Bai WY, Chen XL, Zhang ZT, Fang X, Weng W, Wang YG, Peng HS (2014) Elastic and wearable wire-shaped lithium-Ion battery with high electrochemical performance. Angew Chem Int Ed 53(30):7864–7869
30. Zhang Y, Bai WY, Ren J, Weng W, Lin HJ, Zhang ZT, Peng HS (2014) Super-stretchy lithium-ion battery based on carbon nanotube fiber. J Mater Chem A 2(29):11054–11059

Chapter 8
Fiber-Shaped Integrated Device

Abstract This chapter mainly focuses on the recent advancements of fiber-shaped devices with integrated functions, i.e., solar energy conversion and electric energy storage. Distinguished from conventional integrated device in a conventional planar form, the fiber-shaped integrated device shares a one-dimensional configuration with fiber electrodes. In the beginning, the working mechanisms of the integrated devices are introduced according to the classification. Then two types of fiber-shaped integrated devices are discussed specified by the energy conversion part, i.e., dye-sensitized solar cell and polymer solar cell which are integrated with electrochemical capacitor in the device. Finally, the perspective for the future development of fiber-shaped integrated devices is given.

8.1 Overview of Integrated Device

The concept of integrated device that integrates the functions of solar energy conversion and electric energy storage was aroused nearly at the same time of the appearance of the solar cells, motivated by the intention of storing the generated electric energy and using it whenever needed. According to the configuration of the device, the integrated device can be classified into two categories: all-in-one devices and assembled devices. As the name indicates, the all-in-one devices realize energy conversion and storage in one device, rather than separating them in two individual parts that are connected in assembled devices. The electrode in all-in-one device imparted the ability to harvest solar light as well as store generated charges. In assembled devices, the energy conversion and storage are conducted by solar cell and capacitor that are connected sharing one electrode. Therefore, the energy transfer from the solar cell to the electrochemical capacitor and output to the external circuit can be achieved by manipulating the connection.

H. Peng, *Fiber-Shaped Energy Harvesting and Storage Devices*, Nanostructure Science and Technology, DOI 10.1007/978-3-662-45744-3_8

8.1.1 All-in-One Device

Strictly speaking, the all-in-one device is more agreeable to the concept of integrated device that intends to store the harvested solar energy in the form of electrochemical energy preferable for application. Therefore, a device where the electrode can both harvest and store energy was proposed. As early as the 1980s, the prototype of the integrated device that converts solar energy and stores the generated electric energy emerged based on the photochemical solar cell, a photosynthetic cell employing two redox systems: reacting with the carriers generated at the surface of the semiconductor and the counter electrode [1]. Figure 8.1 shows the configuration of the integrated device. In the photocharging process, the n-type semiconductor was illuminated with radiation higher than its bandgap energy, which induces charge separation within the space charge layer of the semiconductor. This process creates holes at the surface of the semiconductor for polysulfide oxidation and drives electron transfer through an external load and also into electrochemical storage by reduction of SnS to Sn. In the dark, the potential drops below the SnS reduction potential which leads to the spontaneous oxidation of Sn. The electrons flow through the external load, and the discharge process conducts.

Under solar radiation of 96.5 mW cm^{-2}, the photoelectrode generated 23 mA cm^{-2} at 0.495 V, and through the load of 1,500 Ω, a discharge voltage of 0.470 V was generated. The solar cell part of this integrated device showed relatively high direct energy conversion efficiency of 11.8 % and the storage efficiency achieved

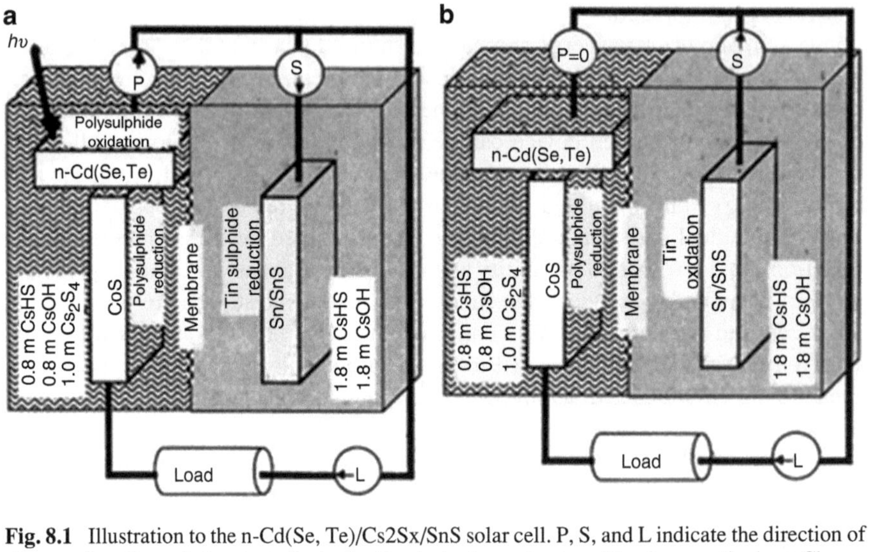

Fig. 8.1 Illustration to the n-Cd(Se, Te)/Cs2Sx/SnS solar cell. P, S, and L indicate the direction of electron flow through the photoelectrode, tin electrode, and external load, respectively. **a** Charge process. **b** Discharge process (Reprinted by permission from Nature Publishing Group Ref. [1], copyright 1987)

95 %. As a result, the overall energy conversion efficiency was 11.2 % calculated by multiplying the direct energy conversion efficiency and storage efficiency, and combining over the load, this value achieved 11.3 %. However, the single crystal-based photoanode requires complex fabrication process and high cost, which calls for easy-fabricated and low-cost solar cells. As a result, dye-sensitized solar cells have been introduced as good candidates to fabricate integrated devices.

As the emergence and prosperity of the dye-sensitized solar cells since 1991, an all-in-one device based on the dye-sensitized solar cell was invented in 2002 [2]. More specifically, this device intended to store the charge excited by the solar light, which performed as a self-charged battery. Generally, a rechargeable battery requires two redox systems transferring electrons and delivering energy at two electrodes. In this case, an extra redox couple involving lithiation and delithiation was needed since the TiO_2 requires a high potential for reduction that is beyond the redox potential of I^-/I_3^-, the settled redox couple as counterpart in the battery. As a result, WO_3 was introduced as the lithium host considering its low lithiation potential (~0 V) and high capability to store lithium ions. The configuration of the self-charged device is displayed in Fig. 8.2, where WO_3, the lithium storage layer, was situated beneath TiO_2, the photoactive layer. Under illumination, the excited dyes injected electrons into the conduction band of TiO_2 which diffused into WO_3. Lithium ions were therefore intercalated into the WO_3 to balance the charge, which meant, in open circuit, the $WO_3|LiWO_3||LiI|LiI_3$ battery was charged. In discharge process, electrons transferred from WO_3 to the counter electrode via an external load and I_3^- ions got electrons and were reduced to I^-. Meanwhile, the intercalated lithium ions were released. Thus, the energy conversion and storage can be carried out in one device. Technically, 0.45 C cm^{-2} can be stored under irradiation of 1,000 W m^{-2} for 1 h, and the open circuit voltage in the charge

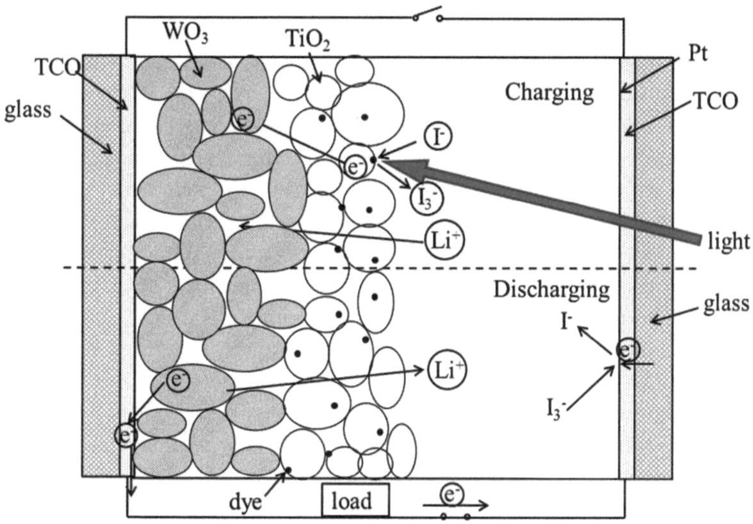

Fig. 8.2 An all-in-one self-charged device based on dye-sensitized solar cell

state is 0.6 V. Since the WO_3 has a high capacitance that can incorporate all of the lithium ions in the electrolyte, the charge capacity was dictated by the lithium concentration. Indeed, elevating the concentration of LiI can increase the charge capacity and suppress the self-discharge but lead to a reduced open circuit voltage. This design enables high charge storage capacity and simple fabrication. However, high storage capacity requires high concentration of LiI, which decreases the open circuit voltage. Moreover, the charge diffusion in the layer of TiO_2 increases the internal resistance of the device and hinders the rapid discharge of the integrated device. To solve the problems, one strategy is to introduce the widely used electrochemical capacitors, which enables a rapid discharge process and high cyclic stability.

8.1.2 Assembled Devices

The progress that assembled devices have made is integrating the professional energy storage device—electrochemical capacitor into the device, instead of the amateur host materials.

8.1.2.1 Two-Electrode System

In 2004, Tsutomu and coworkers reported the first integrated device assembling a dye-sensitized solar cell and an electrochemical capacitor, to store the solar energy in electrochemical capacitor [3]. In this view, this kind of integrated device is also called "photocapacitor." Typical configuration of the device with two electrodes is illustrated in Fig. 8.3a. The assembled integrated device is constructed on multilayered electrodes comprising dye-sensitized semiconductor nanoparticles, hole-trapping layer, and activated carbon particles in contact with an organic electrolyte solution.

In principle, the charge process starts with the light-induced charge separation of dye molecules, and the generated photoelectrons are injected to the conduction band of the semiconductor, which follows the same mechanism of the dye-sensitized solar cell. After charge separation, electrons and holes transfer to activated carbon layers at the counter electrode and photoanode. Positive and negative charges are accumulated on the porous surface of activated carbon that forms the electric double layer in an organic electrolyte with high ionic concentration, and during discharge process in the dark, the stored charges can be used to supply power.

The resulting integrated device showed a capacitance of 0.69 F cm^{-2} and good cyclic stability that during 10 times of charge and discharge cycle, the discharge capacity retained about 85 %. However, the device suffered a stagnant discharge process where electrons have to go through the TiO_2 layer before reaching the electrode, leading to a high internal resistance.

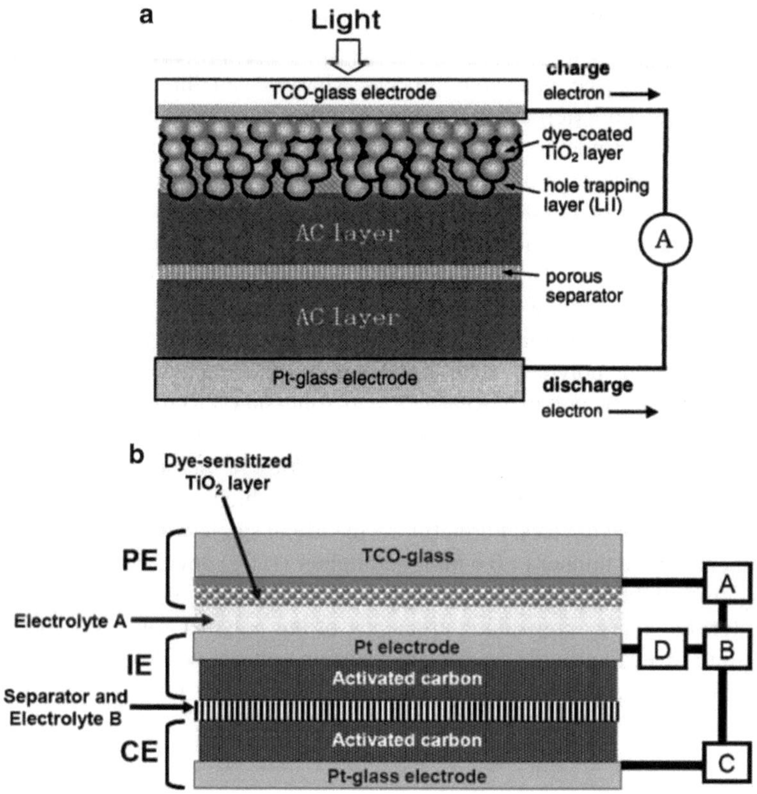

Fig. 8.3 Configurations of the assembled integrated devices. **a** A two-electrode system. **b** A three-electrode system comprising a photoelectrode (PE), an internal electrode (IE), and a counter electrode (CE) (© [2002] IEEE Reprinted from AIP Publishing LLC 2004, with permission, from Ref. [3])

8.1.2.2 Three-Electrode System

In 2005, Tsutomu and coworker modified their two-electrode integrated device by introducing a dual-functional internal electrode [4]. In comparison, the three-electrode system where the energy conversion unit and storage unit share one electrode seems more favorable for electron transfer. As shown in Fig. 8.3b, the three-electrode device comprised a dye-sensitized TiO_2 layer on a transparent conducting glass as photoanode, an activated carbon layer coated on one side of a platinum plate as internal electrode, activated carbon layer coated on the platinum-spattered conducting glass as counter electrode, and two kinds of electrolytes: an electrolyte containing a redox couple of I^-/I_3^- for the dye-sensitized solar cell and an electrolyte for the electrochemical capacitor. The introduced internal electrode, which was sandwiched between the two units, catalyzed the redox reaction in the dye-sensitized solar cell unit and stored charges at the electrochemical capacitor part (Fig. 8.3b).

Compared with the two-electrode configuration, the internal resistance of the integrated device was significantly decreased by the introduction of the internal electrode. During the discharge process, charges can directly transfer to the platinum plate and external circuit without going through the TiO_2 layer. Based on this three-electrode configuration, the resulting integrated device achieved a high charge-state voltage of 0.8 V and large energy output that is five times larger than the two-electrode system. This three-electrode configuration has been widely adopted as the mainstream structure to fabricate integrated devices, and a variety of advancements towards the optimizations of the performance, durability, and processing technique have been achieved since then [5–8].

For example, a printable, all-solid-state integrated device has been fabricated by integrating a polymer solar cell and an electrochemical capacitor [5]. The introduction of polymer solar cell and solid-state electrolyte of the electrochemical capacitor enables an all-solid-state device that increases the stability of the integrated device during practical use. In addition, the layered architecture is compatible with the roll-to-roll printing process, which creates opportunities for printable integrated devices. Furthermore, the use of single-walled carbon nanotube network enables a thinner (<0.6 mm) and lighter (<1 g) device. It should be worth noting that to fully utilize the energy storage capacity of the electrochemical capacitor, the energy conversion efficiency of the polymer solar cell needs to be further improved to provide a sufficient charging voltage, which can be resolved by connecting several solar cells in series.

8.1.3 Materials and Characterization

Nowadays, integrated devices based on electrochemical capacitors become the mainstream, and the material choices are similar with those of dye-sensitized solar cell, polymer solar cell, and supercapacitors, which can be referenced in Chaps. 3, 4, and 6, respectively.

Apart from the basic parameters such as short-circuit current density (J_{SC}), open-circuit voltage (V_{OC}), fill factor (FF) and energy conversion efficiency to evaluate solar cells and specific capacitance, Coulombic efficiency, and cyclic stability to index electrochemical capacitor, a very important parameter that determines the performance of an integrated device is the overall energy conversion efficiency, which can be calculated by dividing the output energy from the electrochemical capacitor by the overall input solar energy.

8.1.4 Summary

Although the energy conversion efficiencies of new-generation photovoltaic devices, dye-sensitized solar cells, and polymer solar cells are still lower than their silicon-based counterparts, they have attracted increasing interest due to a moderate fabrication process with low cost and flexible structure. Their integration with various electrochemical storage devices has been realized with high performance. As a comparison, the generated electric energy from silicon-based solar cells is transported to storage devices through external electric wires with low efficiencies and complex processes. The integrated devices generally appeared in a planar shape, which may limit their practical applications as electronic devices are required to be lighter and smaller in the future. To this end, a fiber-shaped integrated device shows some unique and promising advantages and will be introduced in the next section.

8.2 Overview of Fiber-Shaped Integrated Device

Although the conventional integrated devices have been investigated for more than 30 years, the first successful attempt of fiber-shaped integrated device has not realized until 2011 [9]. Wang and coworkers reported the first fiber-shaped integrated device that combined a dye-sensitized solar cell, a nanogenerator, and an electrochemical capacitor to realize the energy conversion and storage in a single fiber-shaped device. The exciting concept and results have encouraged more efforts to fabricate fiber-shaped integrated devices, and various improvements have been achieved in the structure and stability performance of the integrated device.

Similar to conventional three-electrode integrated device, fiber-shaped device generally shares at least one electrode for the solar cell and electrochemical capacitor. On this account, materials for the shared electrode have to be suitable for both dye-sensitized solar cell and electrochemical capacitor, and various materials such as titanium oxide, zinc oxide, carbon nanotube, and graphene have been investigated. For the solar cell part, both fiber-shaped dye-sensitized solar cell and polymer solar cell have been used as the energy conversion part of the integrated devices. In structure, the fiber-shaped integrated devices are mainly fabricated into coaxial or twisted structure, and the differences in structure require different materials and processing technique choices. Quasi-solid-state and all-solid-state fiber-shaped integrated devices have also been fabricated to improve the durability. To further satisfy the requirement of wearable electronics, stretchable wire-shaped integrated devices have been realized.

8.3 Integrated Devices Based on Dye-Sensitized Solar Cell and Electrochemical Capacitor

Dye-sensitized solar cells (DSCs), as the third-generation photovoltaic cells, are becoming a rising successor of silicon solar cells on the market, benefited from its high energy conversion efficiency, easily fabricating process, and low cost.

For this reason, DSCs are the ideal energy harvesting unit in integrated devices. Integrated devices based on DSCs have been widely studied. Most of the integrated devices are conventional planar structure, which have restrictions in flexibility and size [3, 4, 7, 10–13]. To this end, fiber-shaped energy devices present unique and unusual advantages, for example, being woven into various electronic textiles by traditional textile industry, flexibility for bending devices, and potential for self-powered systems.

As we have discussed in Chap. 4, fiber-shaped DSCs have achieved the highest energy conversion efficiency of 8.45 %, which is favorable for an overall efficiency of the integrated device. The open-circuit voltage in fiber-shaped DSCs can reach ~0.73 V, which suffice for charging the capacitor. Herein, the fiber-shaped integrated devices based on DSCs are discussed.

Electrochemical capacitors as an energy storage device have gained increasing popularity in recent years, due to high power density, fast charge and discharge process, as well as excellent cycle performance. In the planar integrated devices, both electrochemical capacitors and batteries played significant roles [12, 14]. However, in the fiber-shaped integrated devices, electrochemical capacitors stand at the leading position because of its easily fabricating process, convenient integration, as well as relatively low working voltage. The fiber-shaped electrochemical capacitors also exhibit light weight and flexibility, especially high electrochemical performance in integrated devices. These advantages make electrochemical capacitors competent as energy storage units in integrated electronic devices. The energy conversion unit and storage unit can be integrated through a coaxial structure and a twisted structure, which will be discussed in the following sections.

8.3.1 Integrated Device in Coaxial Structure

In coaxial structure, the solar cell unit and electrochemical capacitor unit share one electrode that serves as substrate, and the other electrode of the two units is wound around the communal electrode, respectively. This structure is inspired by planar solar cells and planar supercapacitors. Rolling up the planar solar cells and electrochemical capacitors comes out the coaxial structure. This simple idea first came up with and realized in fiber-shaped integrated devices.

A multifunctional device integrating solar cell, nanogenerator, and electrochemical capacitor in a coaxial structure was fabricated in 2011. Such self-powered fiber system gathered both mechanical and solar energy and then stored in a fiber-shaped electrochemical capacitor (Fig. 8.4). It was the first attempt to integrate fiber-shaped electronic devices and showed the possibility of building up self-powered fiber. The self-powered system shared the Au-modified Kevlar fibers, with ZnO nanowires perpendicularly grown on the surface. Graphene served as the other electrodes of the three units. The energy conversion efficiency is only 0.02 %. And the electrochemical capacitor exhibited capacitance of 0.4 mF cm^{-2} (~0.025 mF cm^{-1}).

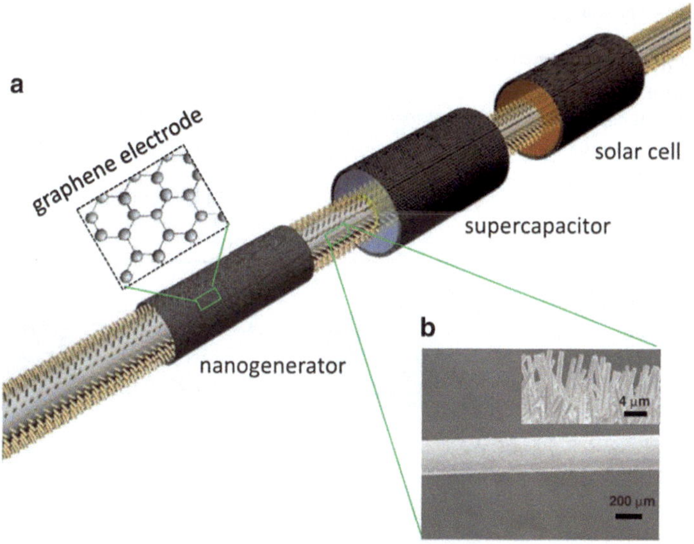

Fig. 8.4 **a** The fiber-shaped self-powered integrated devices comprising a nanogenerator, electrochemical capacitor, and solar cell. **b** The units are assembled in a coaxial structure (Reproduced from Ref. [9] by permission of John Wiley & Sons Ltd)

This integrated device exploited the versatility of electrode materials. Graphene shows high conductivity, transparency, and high specific area, and ZnO nanowire-modified Kevlar fibers exhibited suitable working function for nanogenerator and acceptable specific area for dye absorption and charge storage. However, the inferior performances exposed some shortcomings of the device. For example, the fabrication process of the devices was complicated involving sophisticated techniques and delicate treatments. Technically, the graphene was not self-standing, and therefore, the copper mesh was necessary as substrate in counter electrode, which diminished the transmission of incident light to the dyes. Moreover, compared with TiO_2, ZnO is more chemically vulnerable. Although indeed the performance of the rudimentary integrated device was far from satisfaction, it raised up the new concept that integrate different fiber-shaped devices into a single device to extend their applications.

Some attempts have been devoted to better fiber-shaped integrated devices. Inherited from the improvement of fiber-shaped solar cell and electrochemical capacitor, as described in Chaps. 4 and 5, an energy wire integrating high-performance dye-sensitized solar cell and electrochemical capacitor was fabricated based on TiO_2 and carbon nanotube sheets (Fig. 8.5) [15]. The dye-sensitized solar cell, the photoconversion unit and electrochemical capacitor, and the energy storage unit shared the communal electrode: a TiO_2-modified Ti wire. Carbon nanotube sheets were used as the other electrodes for both units. The device was fabricated by separately winding the carbon nanotube sheet around the modified Ti wire. The TiO_2 nanotubes in photoconversion unit were sensitized by dye N719.

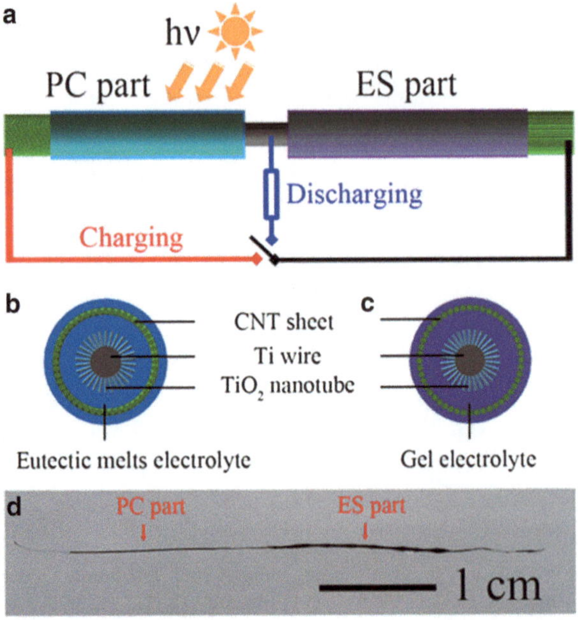

Fig. 8.5 The fiber-shaped coaxial energy wire integrating dye-sensitized solar cell with electrochemical capacitor. **a** Schematic illustration. **b** and **c** Cross-sectional views of the photoconversion and energy storage units of the integrated device, respectively. **d** Photograph of a fiber-shaped integrated device (Reproduced from Ref. [15] by permission of The Royal Society of Chemistry)

Both units used gel electrolyte that have been discussed in Chaps. 4 and 5. The maximal photoelectric conversion efficiency achieved 2.73 %, while the energy storage efficiency reached 75.7 % with specific capacitances up to 0.156 mF cm^{-1} or 3.32 mF cm^{-2} and power densities up to 0.013 mW cm^{-1} or 0.27 mW cm^{-2}. The photoelectric conversion efficiency and the specific capacitances were much higher than the previous research.

Mechanical stability and thermal stability is an essential quality to evaluate integrated devices. Electrolyte is generally the vulnerability of a device which dictates the stability and environmental compatibility. Liquid electrolyte has the advantage of good wettability with two electrodes, but it needs special techniques for sealing, and the volatile liquid delimits the working temperature within its liquid-phase window. In light of the vulnerability of liquid electrolyte, solid-state electrolyte seems more suitable for application and more adaptable for deformation. The endurability to deformation was demonstrated by bending tests. The entire efficiency was maintained by 88.2 % after bending for 1,000 cycles. In addition, the overall efficiency of the integrated device had been maintained by 90.6 % after leaving for 1,000 h, suggesting a decent stability. The coaxial structure is found beneficial for stability. Compared with twisting structure, where the two electrodes are intertwined, the entire device is integrated in one single fiber; bending and knotting the device will not set apart the two electrodes and affect the connection between electrodes and electrolyte. Moreover, in coaxial structure, the counter electrode is expanded around the working electrode, promising a thorough infiltration with the electrolyte and reducing the internal resistances. The coaxial structure is conducive to keep the integrated device intact.

8.3.2 Integrated Device in a Twisting Structure

Twisting structure is popularized in fiber-shaped dye-sensitized solar cells and electrochemical capacitors. It is easily fabricated and convenient to connect with external circuit. Meanwhile, the device performance is more process dependent since the screw pitch and closeness when twisted exert a strong impact on the charge transfer at the electrode interphase, which requires delicate craft to ensure a good performance and replication.

The prototype of fiber-shaped integrated device initially emerged in the wake of the creation of fiber-shaped dye-sensitized solar cell and electrochemical capacitor based on carbon nanotube fibers and TiO_2-modified titanium wires [16]. The photoelectric conversion and energy storage units share a TiO_2-modified titanium wire acting as communal electrode. Here, the aligned titania nanotubes not only improve the charge separation and transport in the photoelectric conversion part but also increase the specific area in the energy storage part. Two individual carbon nanotube fibers were separately wrapped with each part, with screw pitches of approximately 1.1 mm for the photoelectric conversion unit and approximately 0.7 mm for the energy storage unit. The energy conversion efficiency was 2.2 % from the photoelectric conversion unit, and a specific capacitance was 0.6 mF cm^{-2} produced from the storage part. The storage unit can be charged rapidly to a voltage which was close to the open-circuit voltage of the photoelectric conversion unit upon light irradiation. The calculated energy storage efficiency is about 68.4 %, and the entire photoelectric conversion and storage efficiency is 1.5 %, which is obtained by multiplying the energy conversion efficiency and the energy storage efficiency. The charging process of the integrated device is exhibited in Fig. 8.6.

The fiber electrode plays a pivotal role in twisting structure. Flexible materials, like carbon nanotube fibers, ensure two electrodes intimately and easily twisted with each other, without generating internal stress and damage in morphology, and hence, the entire twisted devices possessed high stability during deformation. In contrast, metal wires such as platinum wire can hardly satisfy efficient devices because of the inferior capability in charge storage. In addition, polymer fibers modified with a conductive layer (e.g., indium tin oxide) on the surface were reluctant to present satisfied performance due to poor mechanical stability. Except for carbon nanotube fibers, thin titanium wires as another electrode are flexible enough to twist with the other electrode and meanwhile sufficient to support the whole device. Thus, the integrated device showed high flexibility and stability, which were eligible for portable devices and energy textiles.

Another integrated device based on modified stainless steels and modified titanium wire was fabricated with higher photoelectric conversion efficiency and specific capacitance (Fig. 8.7) [17]. Namely, the titanium wire was coated with a layer of TiO_2 nanoparticles. Stainless steel deposited with polyaniline film served as one electrode of electrochemical capacitor and counter electrode of DSCs. The photoelectric conversion efficiency is 5.41 % based on liquid-based electrolyte, and the overall energy conversion efficiency is up to 2.1 %.

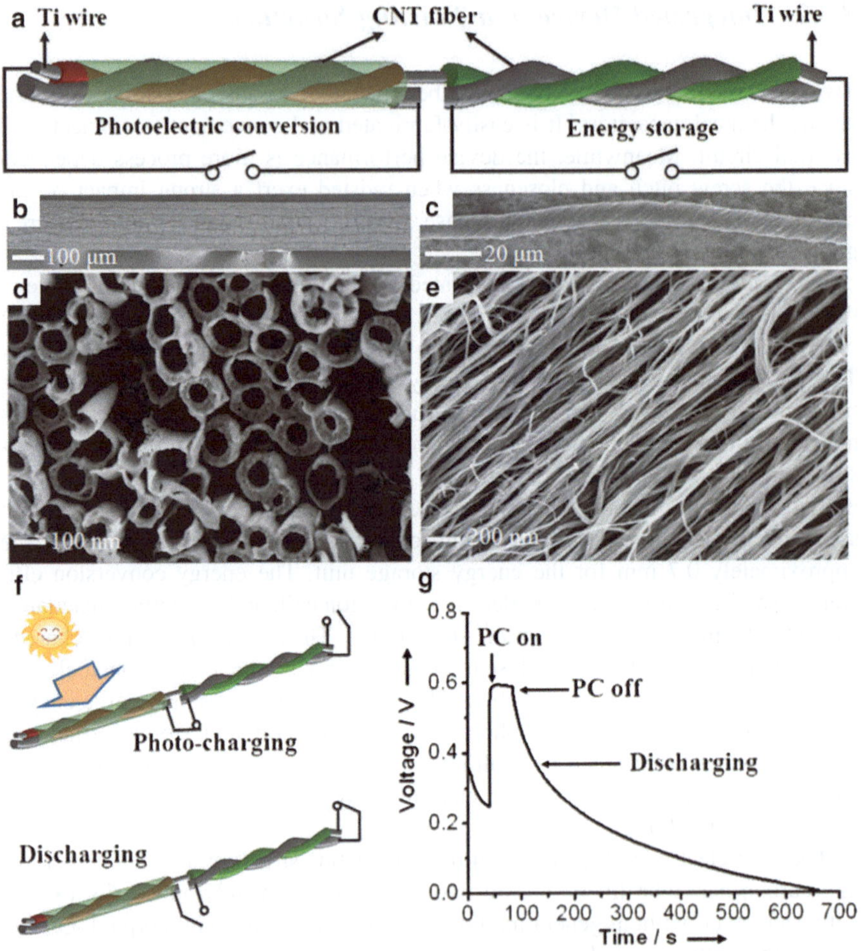

Fig. 8.6 Fiber-shaped integrated device based on dye-sensitized solar cell and electrochemical capacitor in a twisting structure. **a** Schematic illustration of the integrated device for photoelectric conversion (PC) and energy storage (ES). **b** and **d** SEM images of the TiO_2-modified Ti wire at low and high magnifications, respectively. **c** and **e** SEM images of a CNT fiber at low and high magnifications, respectively. **f** Schematic illustration of the circuit connection during charging and discharging processes. **g** Photocharging–discharging curve of a typical energy wire. Discharge current, 0.1 mA (Reproduced from Ref. [16] by permission of John Wiley & Sons Ltd)

Except carbon materials, conductive polymers, such as polyaniline, polythiophene, and polypyrrole, exhibit great potential and excellent performance as electrode materials in integrated devices. These conductive polymers are known for their pseudo-capacitance as active materials in electrochemical capacitors. Meanwhile, conductive polymers, acting as counter electrode, were comparable with platinum electrode in DSCs. Thus, the performance was significantly bettered in both photoelectric conversion and energy storage units.

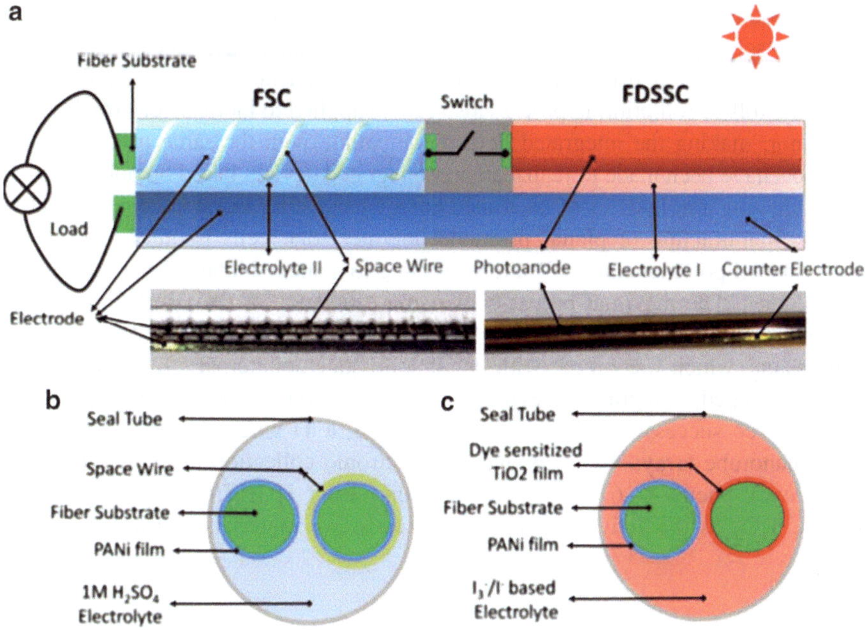

Fig. 8.7 Fiber-shaped integrated device based on dye-sensitized solar cell and electrochemical capacitor based on a twisting structure. **a** Schematic illustration and photographs to the fiber-shaped integrated device. **b**, **c** Cross-sectional views of the photoelectric conversion and energy storage units, respectively (Reproduced from Ref. [17] by permission of The Royal Society of Chemistry)

In a twisting structure, close twisted electrodes are always preferable to ensure a rapid kinetics. However, it arises a problem that plagues the integrated devices, as it have annoyed the dye-sensitized solar cell and electrochemical capacitors, the localized short circuit that leads to severe self-discharge, not least when the device is deformed. The back transfer of electrons can be alleviated when applied with gel electrolyte or introduced separator.

8.4 Integrated Polymer Solar Cell and Electrochemical Capacitor

As another promising branch of photovoltaic devices, polymer solar cells are gaining increasing popularity due to their all-solid-state configuration, where the ionic conductor, liquid electrolyte, is replaced by the hole transport layer and electron transport layer. The fabrication of polymer solar cell, compared with its rival, dye-sensitized solar cell, is carried out in a moderate condition and can be scaled up and massively produced through a printable manufacture and roll-to-roll process. The need for all-solid-state integrated devices naturally motivates the attempt that integrates polymer solar cells with flexible electronic circuits and other electronic devices, such as lithium ion battery and electrochemical capacitors [5, 18]. At the same time, the roll-to-roll printing process also can be applied to fabricate electrochemical capacitor, making the integrated device compatible with the same process. Some attempts have been made to realize planar integrated devices based on polymer solar cells, creating a new series of integrated energy harvesting and storage devices [19].

The manufacture of planar printable polymer solar cells is quite mature; the fiber-shaped polymer solar cells are struggling to achieve higher power conversion efficiencies. The maximal power conversion efficiency of the latest fiber-shaped polymer solar cell was 3.87 %, and the open-circuit voltage reached ~0.6 V in a wire shape, which was on par with its all-solid-state counterparts [20]. At present, a fiber-shaped integrated device based on polymer solar cell has been materialized successfully. As shown in Fig. 8.8, a Ti wire vertically grown with TiO_2 nanotube arrays was employed as electronic collection layer. Poly(3-hexyl thiophene):-phenyl-C 61-butyric acid methyl ester (P3HT:PCBM) was coated on the TiO_2 as the active materials. Outside active materials, the hole transport

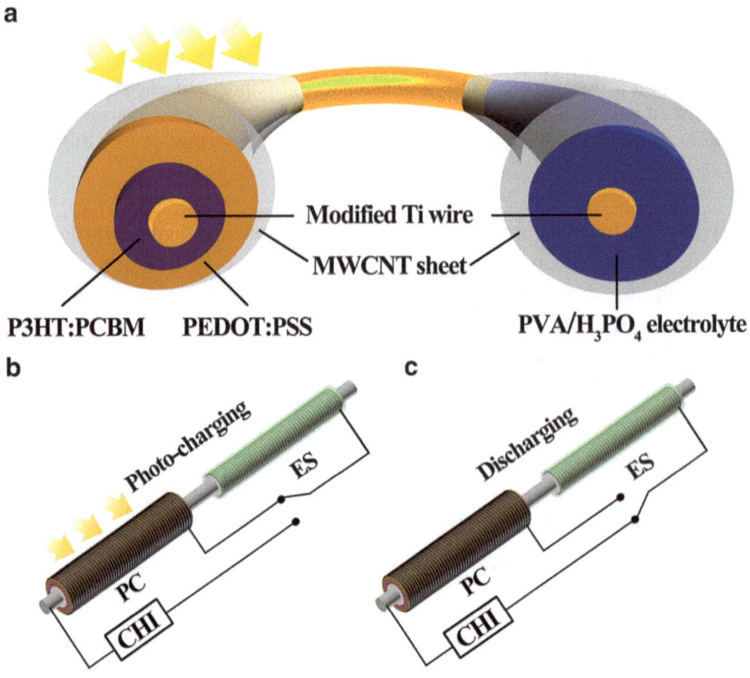

Fig. 8.8 Fiber-shaped integrated device based on polymer solar cell and electrochemical capacitor. **a** Schematic illustration to the fiber-shaped integrated device. The left and right sections correspond to the photoelectric conversion and energy storage units, respectively. **b** and **c** The circuit connection in charge and discharge, respectively (Reproduced from Ref. [19] by permission of John Wiley & Sons Ltd)

material was coated via dip-coating method. Note that TiO_2 nanotube arrays were only 1.8 μm in height to ensure a short path for electron and hole transport. Carbon nanotube sheets were wound as the counter electrode. The energy storage unit was fabricated following the same methods presented in last section.

This device delivered satisfactory performance in both polymer solar cell and electrochemical capacitor. The efficiency of polymer solar cells reached 1.01 %, and the specific capacitance in length is 0.077 mF cm^{-1}. The overall efficiency of the device is 0.82 %. All-solid-state feature imparts the integrated devices with extraordinary stability during bending deformation. In the absence of liquid electrolyte, the integrated device can be deformed into various shapes adapting to the application requirement without deterioration in performance. The entire photoelectric conversion and storage efficiency was slightly decreased by less than 10 % after bending 1,000 cycles.

All-solid-state fiber-shaped polymer solar cells were more ideal independent modules for integration than planar polymer solar cells. In the planar form products integrated with polymer solar cells, the polymer solar cells firstly set as an independent module, followed by printing electronic circuitry on top of the solar cell and semiautomatic adding of discrete components to meet the application requirements [18]. The all-solid-state fiber-shaped polymer solar cells or integrated devices are much easier to insert into electronic circuitry and meet the demand of precise circuit design and even can be woven into with each other or with other chemical fibers to form flexible textiles. The refined designed energy textiles or meticulous circuit can output different voltages through series and parallel for adjusting to various applications.

There is however a significant difference between simply testing an integrated device in the laboratory and testing a product into the hands of the user. So far only a few products achieved the latter demand, such as a flexible simple lamp for the Lighting Africa initiative [21]. None of commercial products are based on fiber-shaped solar cells, which indicated plenty of difficulties need to be resolved and potential applications remain to be developed.

8.5 Stretchable Fiber-Shaped Integrated Device

Apart from the improvements on the structure and performance of the integrated devices, another important aspect pertaining to wearable applications is the stretchability of fiber-shaped integrated devices. As a result, to develop stretchable fiber-shaped integrated device is highly important to meet the requirement of wearable applications, and yet few achievements have been made towards this direction [22].

One approach to make the integrated device stretchable was integrating stretchable dye-sensitized solar cell and electrochemical capacitor with a coaxial structure. Specifically, a polymer tube was set between the photoanode and electrochemical capacitor for preventing electrolyte infiltrating into electrochemical capacitor,

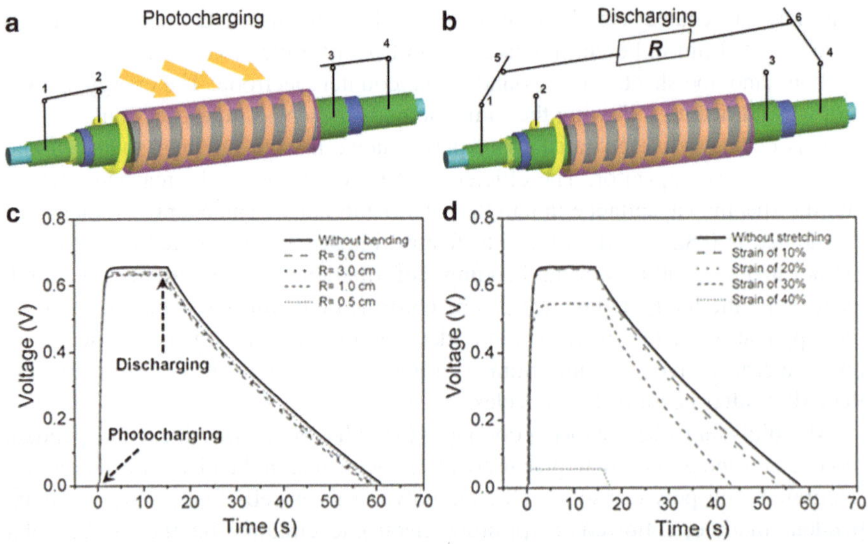

Fig. 8.9 Stretchable fiber-shaped integrated device based on dye-sensitized solar cell and electrochemical capacitor. **a** and **b** Circuit connection in photocharge and discharge process, respectively. **c** Photocharge and discharge processes without and with bending with curvature radius of 5.0, 3.0, 1.0, and 0.5 cm, respectively. **d** Photocharge and discharge processes before and after stretching with strains of 10 %, 20 %, 30 %, and 40 %, respectively. The galvanostatic discharging process was performed by an electrochemical station at a current density of 0.1 A g^{-1} (Reproduced from Ref. [22] by permission of John Wiley & Sons Ltd)

and layers of aligned carbon nanotube sheets were wrapped on the separation tube acting as the counter electrode. At the outermost part, another polymer separation tube was applied to holding the liquid electrolyte and protecting the whole device. Actually the configuration is inspired by the planar photo-supercapacitor [14]. Simply rolling up a typical planar photo-supercapacitor came a real coaxial fiber-shaped integrated device. The photocharge and discharge process for the self-powering energy fiber was displayed in Fig. 8.9.

The stability of integrated devices is a critical performance. In practical application, not only bending property but stretching property has great influence on performance of integrated devices. Owing to the stretchability of electrochemical capacitor and the spring shape of photoanode, the whole integrated device keeps stable during the deformation. The entire energy conversion and storage efficiency was calculated to be 1.83 % and can be well maintained after stretching.

Compared with the traditional integrated devices which were designed by fabricating the solar cells and electrochemical capacitor at the two ends of a fiber, this kind of configuration overcomes the difficulty of extracting electrodes to connect external circuit and switching from charge and discharge. In addition, the configuration overcame the fragility in connecting section. Furthermore, the light utilization efficiency in unit area was enhanced on account of inserting the electrochemical capacitors into the solar cell. All the above reveal the extraordinary advantages of stretchable fiber-shaped integrated devices over other devices.

8.6 Perspective

With a novel one-dimensional configuration, fiber-shaped integrated devices share the advantages of light weight, flexibility, weavability, and wearability, which hold great potential as miniature self-powered devices in the future. Although the investigation on fiber-shaped integrated devices has just started, this field is growing rapidly, and more advancement will be achieved to further improve the structure, stability, and performance of the integrated device.

The overall energy conversion efficiency is the most important factor that determined the energy conversion and storage performance of an integrated device. Up to date, the highest overall energy conversion efficiency has achieved 11.2 %, which needs to be further increased. It is well recognized that increasing the energy conversion efficiency is the most efficient method to increase the overall energy conversion efficiency. As a result, to develop solar cells with higher energy conversion efficiency is critical for the further improvement of the fiber-shaped integrated devices.

Electrolyte is another critical problem that can be further improved in fiber-shaped integrated devices. The energy conversion efficiency of dye-sensitized solar cells using liquid electrolyte is generally higher than that uses quasi-solid-state or all-solid-state electrolyte, which is contributable to a higher overall energy conversion efficiency of the whole device. However, a fiber-shaped integrated device using liquid electrolyte needs strict and complex sealing processes to avoid leaking during use. In addition, the operation temperature cannot exceed 70 °C; otherwise, the liquid electrolyte will evaporate and the integrated device deteriorates. Quasi-solid-state and all-solid-state electrolytes can overcome the problems above, but the resulting energy conversion efficiency is generally low. As a result, to develop better quasi-solid-state and all-solid-state electrolytes for high-performance fiber-shaped dye-sensitized solar cell is necessary. On the other hand, developing high-performance fiber-shaped polymer solar cells is also preferred, though the present energy conversion efficiency is still much lower than fiber-shaped dye-sensitized solar cell.

The scale-up fabrication is another important issue for the practical applications of the fiber-shaped integrated devices, while no efforts have been made towards this direction. The present lengths of the integrated devices are all on the level of centimeters, which are obviously too short for practical applications. With the extension of integrated device length, some new problem will come up, e.g., the internal resistance may increase due to the extension offiber-shaped electrode, which will severely decrease the performance of the whole device. As a result, the resistance of the fiber-shaped electrodes should be taken into consideration to meet the length requirement in future applications.

References

1. Licht S, Hodes G, Tenne R, Manassen J (1987) A light-variation insensitive high-efficiency solar cell. Nature 326(6116):863–864
2. Hauch A, Georg A, Krasovec UO, Orel B (2002) Photovoltaically self-charging battery. J Electrochem Soc 149(9):A1208–A1211
3. Miyasaka T, Murakami TN (2004) The photocapacitor: an efficient self-charging capacitor for direct storage of solar energy. Appl Phys Lett 85(17):3932–3934
4. Murakami TN, Kawashima N, Miyasaka T (2005) A high-voltage dye-sensitized photocapacitor of a three-electrode system. Chem Commun 26:3346–3348
5. Wee G, Salim T, Lam YM, Mhaisalkar SG, Srinivasan M (2011) Printable photo-supercapacitor using single-walled carbon nanotubes. Energy Environ Sci 4(2):413–416
6. Hsu CY, Chen HW, Lee KM, Hu CW, Ho KC (2010) A dye-sensitized photo-supercapacitor based on PProDOT-Et-2 thick films. J Power Sources 195(18):6232–6238
7. Chen HW, Hsu CY, Chen JG, Lee KM, Wang CC, Huang KC, Ho KC (2010) Plastic dye-sensitized photo-supercapacitor using electrophoretic deposition and compression methods. J Power Sources 195(18):6225–6231
8. Xu J, Wu H, Lu LF, Leung SF, Chen D, Chen XY, Fan ZY, Shen GZ, Li DD (2014) Integrated photo-supercapacitor based on bi-polar TiO_2 nanotube arrays with selective one-sideplasma-assisted hydrogenation. Adv Funct Mater 24(13):1840–1846
9. Bae J, Park YJ, Lee M, Cha SN, Choi YJ, Lee CS, Kim JM, Wang ZL (2011) Single-fiber-based hybridization of energy converters and storage units using graphene as electrodes. Adv Mater 23(30):3446–3449
10. Zhang X, Huang X, Li C, Jiang H (2013) Dye-sensitized solar cell with energy storage function through PVDF/ZnO nanocomposite counter electrode. Adv Mater 25(30):4093–4096
11. Miyasaka T, Ikeda N, Murakami TN, Teshima K (2007) Light energy conversion and storage with soft carbonaceous materials that solidify mesoscopic electrochemical interfaces. Chem Lett 36(4):480–487
12. Weng Z, Guo H, Liu X, Wu S, Yeung K, Chu PK (2013) Nanostructured TiO_2 for energy conversion and storage. RSC Adv 3(47):24758–24775
13. Skunik-Nuckowska M, Grzejszczyk K, Kulesza PJ, Yang L, Vlachopoulos N, Häggman L, Johansson E, Hagfeldt A (2013) Integration of solid-state dye-sensitized solar cell with metal oxide charge storage material into photoelectrochemical capacitor. J Power Sources 234:91–99
14. Guo W, Xue X, Wang S, Lin C, Wang ZL (2012) An integrated power pack of dye-sensitized solar cell and Li battery based on double-sided TiO_2 nanotube arrays. Nano Lett 12(5):2520–2523
15. Chen X, Sun H, Yang Z, Guan G, Zhang Z, Qiu L, Peng H (2014) A novel "energy fiber" by coaxially integrating dye-sensitized solar cell and electrochemical capacitor. J Mater Chem A 2(6):1897–1902
16. Chen T, Qiu L, Yang Z, Cai Z, Ren J, Li H, Lin H, Sun X, Peng H (2012) An integrated "energy wire" for both photoelectric conversion and energy storage. Angew Chem Int Ed 51(48):11977–11980
17. Fu Y, Wu H, Ye S, Cai X, Yu X, Hou S, Kafafy H, Zou D (2013) Integrated power fiber for energy conversion and storage. Energy EnvironSci 6(3):805–812
18. Krebs FC, Fyenbo J, Jørgensen M (2010) Product integration of compact roll-to-roll processed polymer solar cell modules: methods and manufacture using flexographic printing, slot-die coating and rotary screen printing. J Mater Chem 20(41):8994–9001
19. Zhang Z, Chen X, Chen P, Guan G, Qiu L, Lin H, Yang Z, Bai W, Luo Y, Peng H (2014) Integrated polymer solar cell and electrochemical supercapacitor in a flexible and stable fiber format. Adv Mater 26(3):466–470

20. Lee MR, Eckert RD, Forberich K, Dennler G, Brabec CJ, Gaudiana RA (2009) Solar power wires based on organic photovoltaic materials. Science 324(5924):232–235
21. Krebs FC, Nielsen TD, Fyenbo J, Wadstrøm M, Pedersen MS (2010) Manufacture, integration and demonstration of polymer solar cells in a lamp for the "Lighting Africa" initiative. Energy EnvironSci 3(5):512–525
22. Yang Z, Deng J, Sun H, Ren J, Pan S, Peng H (2014) Self-powered energy fiber: energy conversion in the sheath and storage in the core. Adv Mater 28(41):7038–7042

Chapter 9
Energy Textiles

Abstract This chapter mainly focuses on energy textile, i.e., photovoltaic textiles and supercapacitor textiles. The fiber-shaped energy device is aimed to satisfy the ever-growing need for wearable electronics. Energy textiles, exemplified by dye-sensitized solar cell fabrics and supercapacitor fabrics, are presented here. Their fabrication, performances, and properties are discussed.

9.1 Overview of Energy Textiles

As pundits advertise the prospects of wearable electronic products, fever in fiber-shaped devices has swept over the world in recent years that renews the interest in miniature, flexible, and high-performance energy fabrics. The conventional energy devices, solar cells, supercapacitors, lithium ion batteries, and piezoelectric generators have been transformed into fiber shape, in the wake of the success in fiber electrodes [1, 2]. Successively, several prototypes of energy textiles have been fabricated in two approaches: weaving the fiber-shaped devices into fabric and assembling the device based on fabric electrodes.

The discussion of energy textiles based on fiber-shaped devices [3–10] is scattered in Chaps. 3, 4, 5, 6, and 7. Here, we will make a summarization. In Chap. 4, we demonstrated a polymer solar cell (PSC) fabric woven from fiber-shaped PSCs (Fig. 9.1a). A dye-sensitized solar cell (DSC) fabric is exhibited in Fig. 9.1b (Chap. 3). Figure 9.1c displays an energy storage fabric made from fiber-shaped chromatic supercapacitors (Chap. 6), while Fig. 9.1d shows the lithium ion battery fabric woven into a cloth (Chap. 7). Though rudimentary, these energy textiles still exemplify the application of the fiber-shaped devices to conform the conception of wearable devices. The prototypes of energy textile demonstrated here are woven from fiber-shaped device units, which are usually based on two electrodes. In light of practical application, the PSC fabric seems more appealing since its all-solid state gets rid of the difficulty in encapsulation and safety vulnerability of liquid electrolyte, which still dominants in fiber-shaped DSCs. Both of the PSC fabric and DSC fabric are constructed in a twisting structure containing modified Ti wire and carbon nanotube (CNT) fiber. As Ti wire is less flexible than CNT fiber,

H. Peng, *Fiber-Shaped Energy Harvesting and Storage Devices*, Nanostructure Science and Technology, DOI 10.1007/978-3-662-45744-3_9

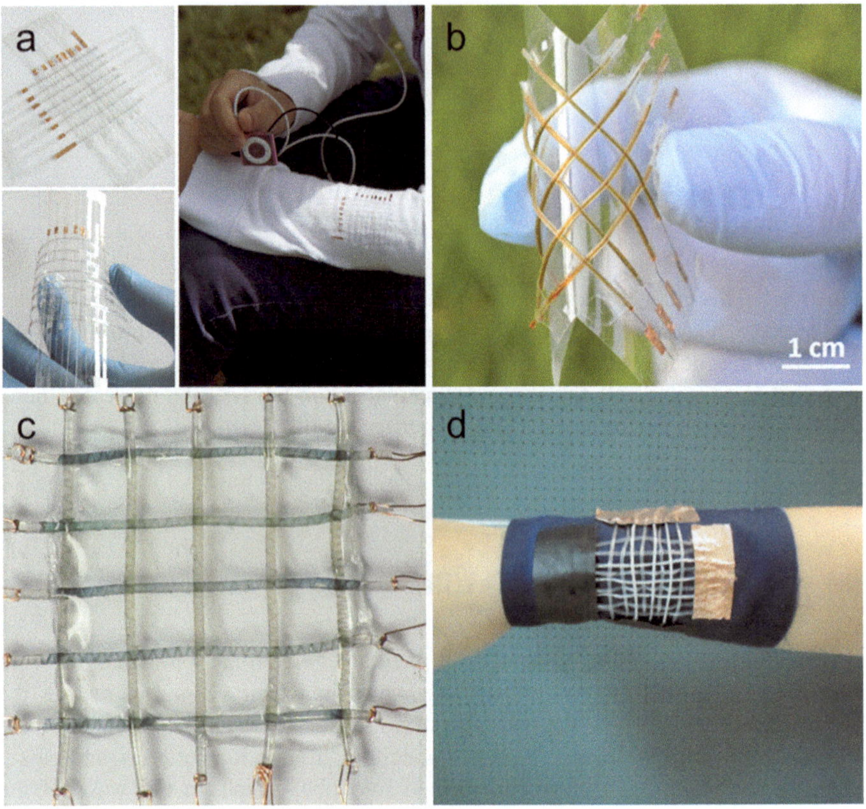

Fig. 9.1 Energy textiles woven from fiber-shaped energy devices. **a** A polymer solar cell (PSC) fabric (Reproduced from Ref. [9] by permission of John Wiley & Sons Ltd). **b** A dye-sensitized solar cell (DSC) fabric (Reproduced from Ref. [10] by permission of John Wiley & Sons Ltd). **c** A chromatic supercapacitor textile (Reproduced from Ref. [11] by permission of John Wiley & Sons Ltd). **d** The lithium ion battery fabric woven into a cloth (Reproduced from Ref. [8] by permission of John Wiley & Sons Ltd)

it is reasonable to speculate that the two electrodes are inclined to slip off when deformed. Gel electrolytes are both adopted in fiber-shaped supercapacitors and lithium ion batteries, making their fabrics more compatible to surroundings. Weaving into fabrics always requires long units, but the elongated units often have a reduced performance.

9.2 Energy Textiles from Fabric Electrode

The contradiction between the length and performance can be figured out through a new strategy: fabricating energy textile with fabric electrode.

9.2.1 Photovoltaic Textiles

Here a DSC fabric can be made from two fabric electrodes. Similar to the fiber-shaped DSCs, dye-sensitized TiO_2 nanotubes are used as photoanode which yet are grown on a fabric woven from Ti wires (130 μm in diameter) by electrochemical anodization. Figure 9.2a shows a scanning electron microscopy (SEM)

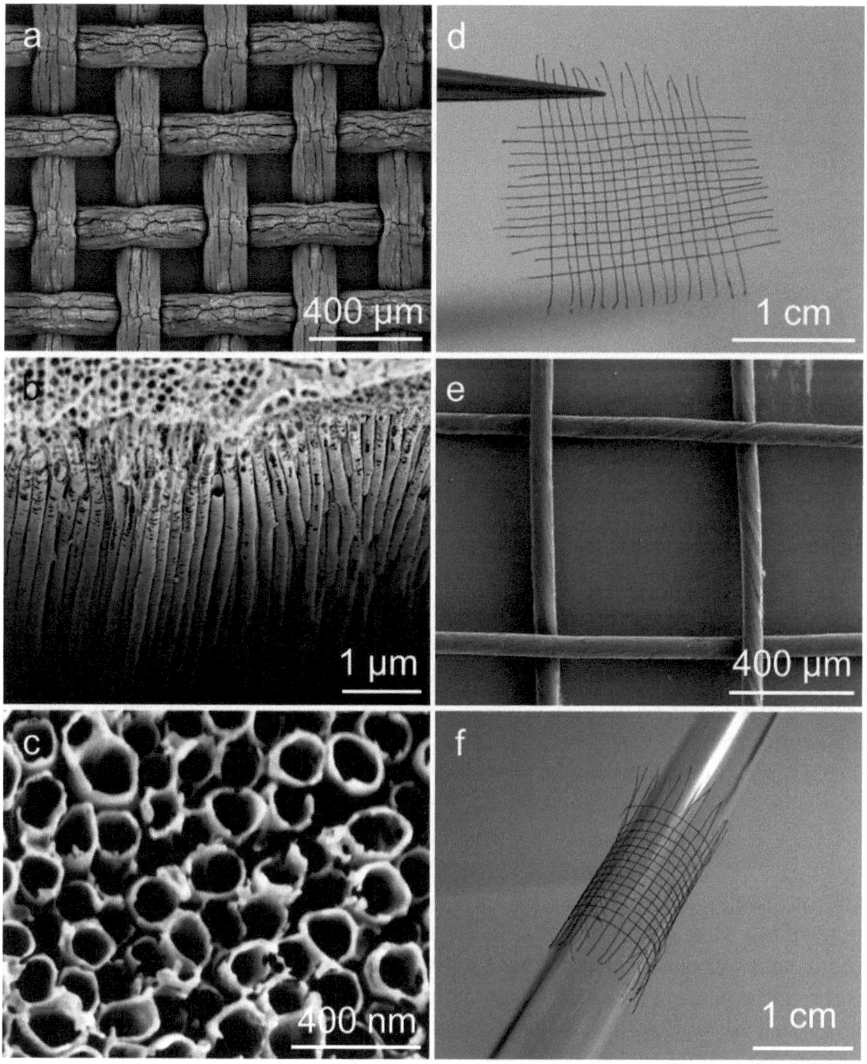

Fig. 9.2 **a** SEM image of the modified Ti wire fabric. **b** and **c** SEM images of TiO_2 nanotubes at low and high magnifications, respectively. **d** and **e** Photograph and SEM image of a CNT fiber fabric, respectively. **f** Photograph of the CNT fiber fabric wrapped around a glass rod (Reproduced from Ref. [7] by permission of John Wiley & Sons Ltd)

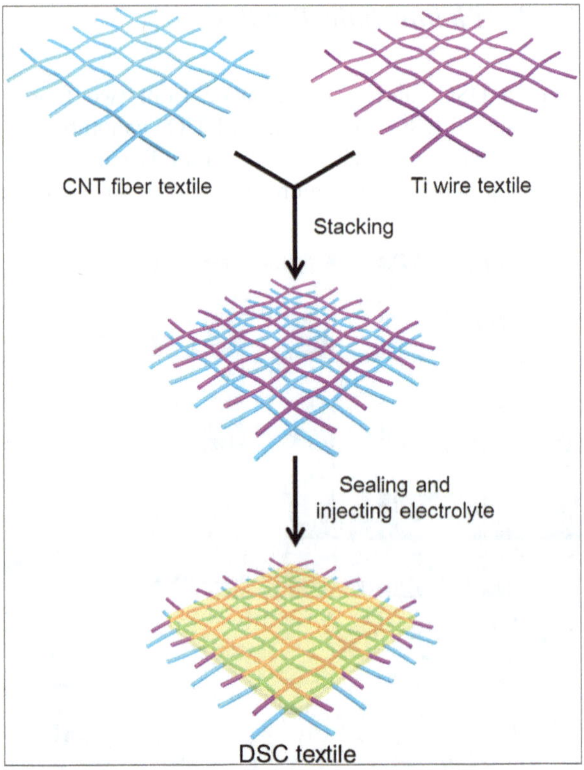

Fig. 9.3 The fabrication of DSC textile form fabric electrodes (Reproduced from ref. [7] by permission of John Wiley & Sons Ltd)

image of a Ti wire fabric after anodizing for 6 h. The diameter of nanotube is 70–100 nm (Figs. 9.2b, c). The length of TiO_2 nanotube can be adjusted with anodizing time [7]. The fabric counter electrode is prepared from CNT fibers with high conductivity (10^2–10^3 S cm^{-1}) and mechanical strength (~600 MPa) (Fig. 9.2d, e). The resulting CNT fiber fabric is free-standing and conformable to different surfaces (Fig. 9.2f).

The DSC fabric fabrication is much easier. As illustrated in Fig. 9.3, the CNT fiber fabric is stacked onto the modified Ti wire fabric, followed by injecting electrolyte. As we have discussed in Chap. 3, the lengths of TiO_2 nanotubes have an impact on cell efficiencies, which reflects the competing effect between the dye absorption and charge recombination. The optimal length is selected by controlled trials in Fig. 9.4a. A DSC fabric with 30 μm of TiO_2 nanotubes delivered the highest energy conversion efficiency (3.10 %), which is in line with our former results. It should be noted that the illuminated area, relating to the current density (J_{SC}), is calculated by multiplying the length and width of the DSC fabric. The interspaces among the wires are included though they do not contribute to photocurrent. Therefore, the obtained apparent efficiency is lower than the fiber-shaped DSC.

The DSC fabric made from fabric electrodes was flexible in structure and sustainable in performances, as validated by the overlapped *J–V* curves before

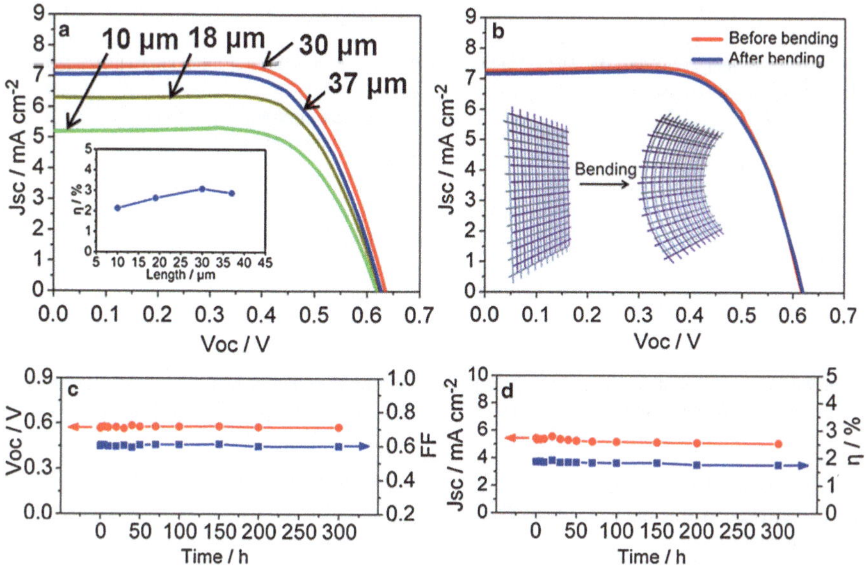

Fig. 9.4 Photovoltaic performances of the DSC fabric. **a** *J–V* curves of the DSC fabric with different lengths of TiO_2 nanotubes (the inserted graph shows the dependence of energy conversion efficiency (η) on length of TiO_2 nanotubes). **b** *J–V* curves of the DSC fabric with the TiO_2 nanotube length of 30 μm before and after bending based on the liquid electrolyte. **c** and **d** Dependence of photovoltaic parameter of the DSC fabric based on the solid electrolyte on time (Reproduced from Ref. [7] by permission of John Wiley & Sons Ltd)

and after bending in Fig. 9.4b. As solid-state electrolyte is more adaptable to environment, displacing the liquid electrolyte with a solid surrogate, the DSC fabric exhibited a better stability. Figure 9.4c, d shows the photovoltaic parameters traced in 300 h exposed in air, the efficiency decayed by less than 6 % during this period.

Sharing the same two-dimensional configurations, the DSC fabric displays a salient superiority over the traditional planar DSCs. In the DSC fabric, the incident light is less shadowed by the grid of counter electrode when back illuminated. In contrast, the shadow and reflection effect from counter electrode and electrolyte often annoy the planar DSC. Their responses to front and back illumination are compared in Fig. 9.5. For the DSC fabric, the responsive photocurrent difference is less than 10 %, while it is half decayed in planar DSC. Moreover, the planar DSC is selective to the direction of incident light. The highest efficiency can be achieved provided that the cell is placed right under the illumination. However, the selection in DSC fabric is alleviated benefited from the fiber-shaped DSC unit which is all around available for photoconversion. As a demonstration, the power conversion efficiency of the DSC fabric is fluctuated within 10 % at various incident lights (Fig. 9.6) [7]. The adaptability to incident direction is of great importance to practical applications. For example, as the incident angle of sunlight varies over the time, the output power of the DSC fabric is less affected than the planar DSC. And the DSC fabric has advantage in taking in the diffused light from the surroundings.

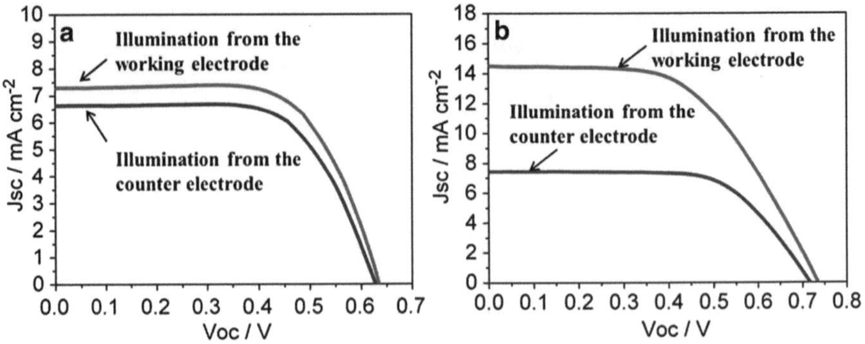

Fig. 9.5 *J-V* curves of a DSC fabric (**a**) and the conventional planar DSC (**b**) under illumination from the two opposite sides (Reproduced from Ref. [7] by permission of John Wiley & Sons Ltd)

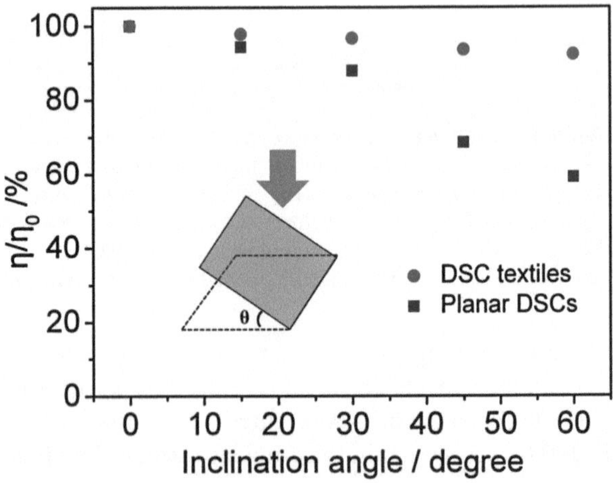

Fig. 9.6 Dependence of power conversion efficiency on the incident light by lifting one edge of a DSC with an angle of (θ). Here η_0 and η correspond to the energy conversion efficiencies at 0 and the other angle, respectively (Reproduced from Ref. [7] by permission of John Wiley & Sons Ltd)

The DSC fabrics can be easily integrated into various flexible substrates. As shown in Fig. 9.7a, a DSC fabric can be integrated into a cloth which is sufficient to power a red light-emitting diode (LED) (Fig. 9.7b).

Polymer solar cell (PSC) textile can be also realized by using the textile electrode. Briefly, the anodized Ti wire textile is deposited with a layer of TiO_2 nanoparticles. Poly(3-hexylthiophene)/phenyl-C_{71}-butyric acid methyl ester (P3HT:PCBM) and poly(3,4-ethylenedioxythiophene)/poly(styrene sulfonate) (PEDOT:PSS), which are served as active layers, are sequentially dip coated onto the modified Ti wire textile. Finally, two flexible, aligned CNT sheets are finally paved onto and adhered to both sides of a Ti wire textile to produce the PSC textile (Fig. 9.8). Due to the sandwiched structure, the PSC textile displays almost the same energy conversion efficiencies at both sides, and the power conversion efficiencies are varied by less than 3 % after bending for over 200 cycles. Interestingly, this PSC textile shows a low areal density of 5.9 mg cm^{-2}, compared with 20 mg cm^{-2} of PET/ITO-based solar cells, 31.3 mg cm^{-2} of flexible film-based PSCs, and 173 mg cm^{-2} of DSC textiles [12].

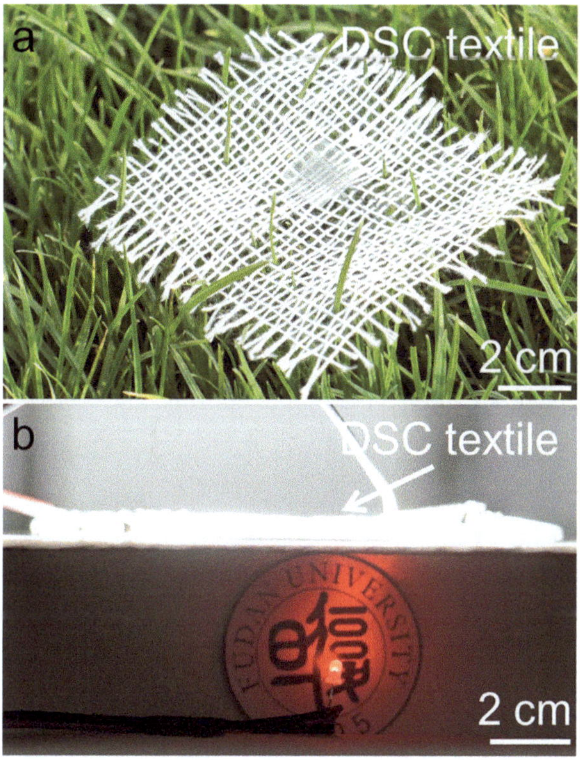

Fig. 9.7 **a** Photograph of a DSC fabric being integrated in a cloth. **b** A red light-emitting diode (LED) was lightened up by a DSC fabric (five in series) (Reproduced from Ref. [7] by permission of John Wiley & Sons Ltd)

9.2.2 Supercapacitor Textiles

Besides the fiber electrode, using the fabrics as electrodes is another effective and simple approach to realize the wearable ability. Recently, fabrics have been used as a component for fabricating the supercapacitor. For instance, Cui and coworkers [13] have designed a flexible fabric supercapacitor, where the solution-based method was used to turn cotton fabric into conductive fabric by coating the cellulose fibers with carbon nanotubes or graphene sheets. Screen-printed activated carbon onto cotton and polyester woven fabrics were also used as fabric electrode for supercapacitor applications. Li and coworkers [14] carbonized cotton fabric into activated carbon fabric for fabricating a hybrid supercapacitor by incorporating MnO_2 particles. Fabrication of supercapacitor textiles from commercial fabrics is simple and feasible. However, it should be recognized that supercapacitors with composite fabrics or activated fabrics are all opaque.

Supercapacitor textiles can also be fabricated by stacking CNT fiber fabric electrodes. In order to enhance the capacity, polyaniline (PANI) is incorporated into the CNT fiber fabric by electrochemical deposition. The content of PANI is monitored by electric quantity. Figures 9.9a–f show typical SEM images of the resulting CNT/PANI fiber fabric with different PANI contents, which lead to

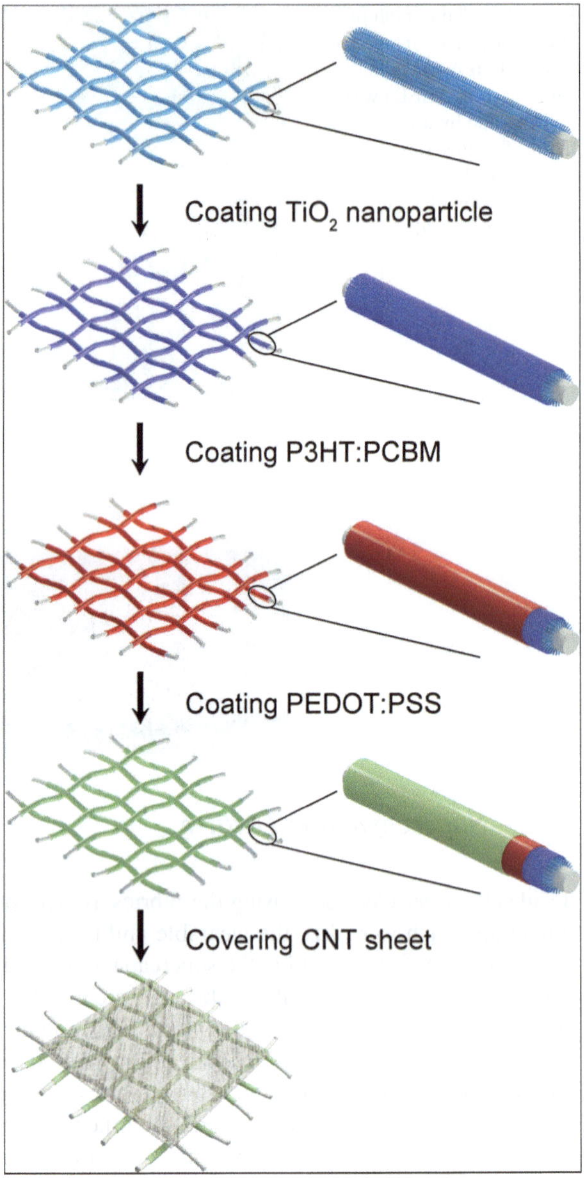

Fig. 9.8 Schematic illustration to the fabrication process of the PSC textile (Reproduced from Ref. [12] by permission of John Wiley & Sons Ltd)

different morphologies. The over deposited PANI tends to aggregate as shown in Fig. 9.9f. The supercapacitor is fabricated by stacking two CNT/PANI fiber fabrics separated by gel electrolyte (Fig. 9.10) [15].

The specific capacitances are increased from 7.7, 108.5, 152.8, 201.8, to 272.7 F g^{-1} with the increasing PANI weight percentage from 0, 15 %, 30 %, 40 %, to 50 %, respectively. When it exceeds 60 %, the specific capacitance is slightly decreased to 240.6 F g^{-1}. Note that the capacitance is specified to the weight

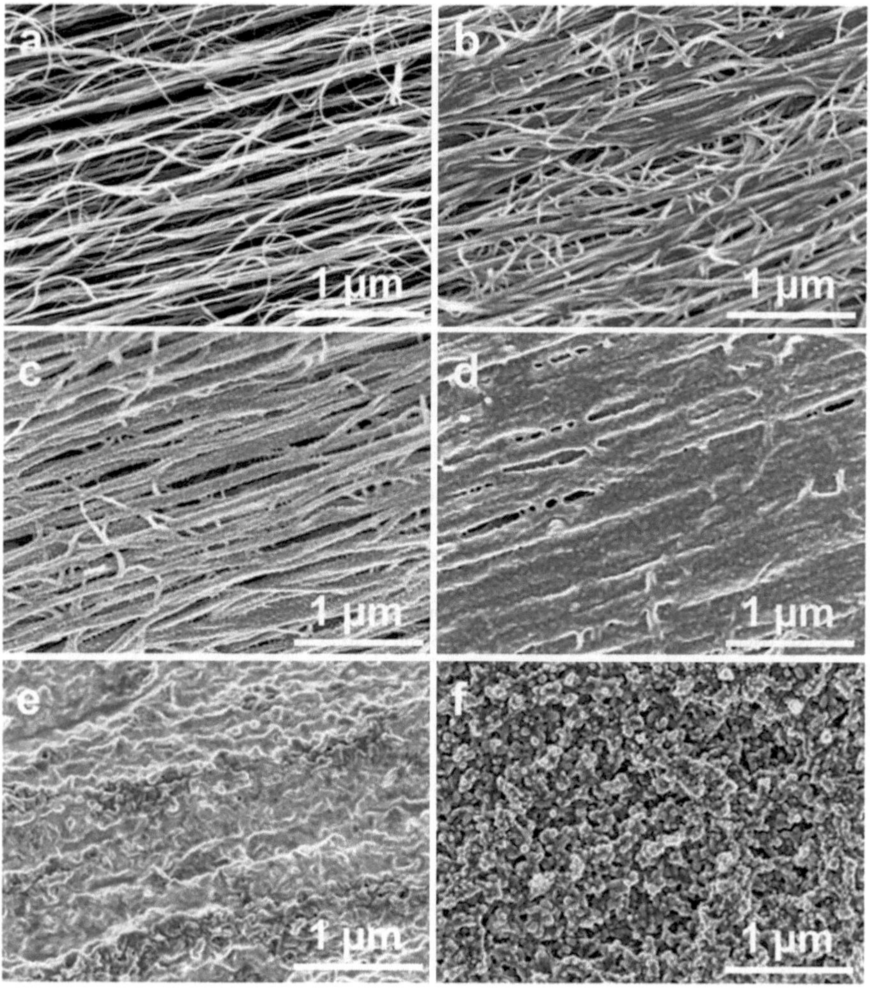

Fig. 9.9 SEM images of CNT fiber fabrics at different PANI weight percentages. **a** A pristine aligned CNT. **b–f** Aligned CNT/PANI composites with PANI weight percentage of 15 %, 30 %, 40 %, 50 %, and 60 %, respectively (Reproduced from Ref. [15] by permission of John Wiley & Sons Ltd)

of the single fabric electrode. Figure 9.11a shows cyclic voltammogram curves whose shapes are well kept at different scan rates. The redox peaks derived from different oxidation states of PANI indicate a pseudocapacitance in the CNT/PANI composite fabric. Figure 9.11b shows the galvanostatic charge/discharge curve with increasing current densities. All the curves show a typical triangular shape, with negligible voltage skip, which indicates that the supercapacitor fabric can stably perform at a wide range of current density. Figure 9.11c shows the cyclic performances of supercapacitor fabrics. As expected, the supercapacitor fabric exhibits a high cyclic stability, e.g., over 90 % in capacitance after 2000 charge–discharge cycles with the Coulombic efficiency 100 % retained (Fig. 9.11d).

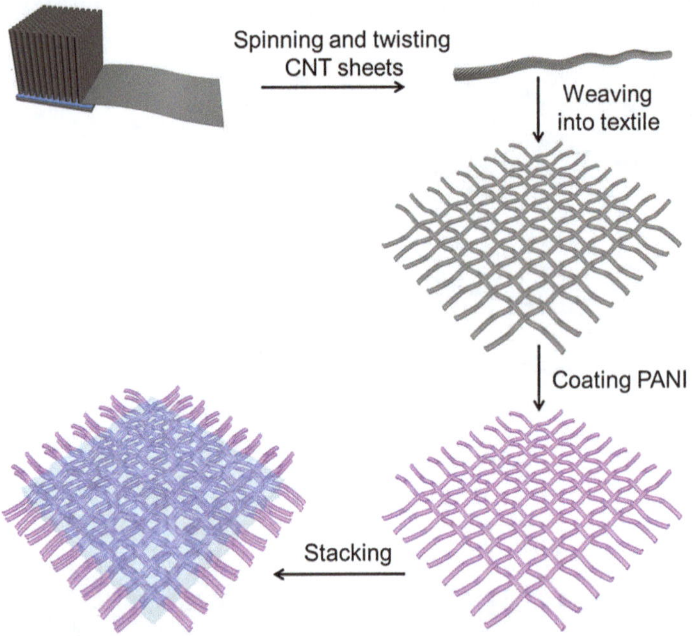

Fig. 9.10 The fabrication of the supercapacitor textiles based on CNT/PANI composite fabric electrode (Reproduced from Ref. [15] by permission of John Wiley & Sons Ltd)

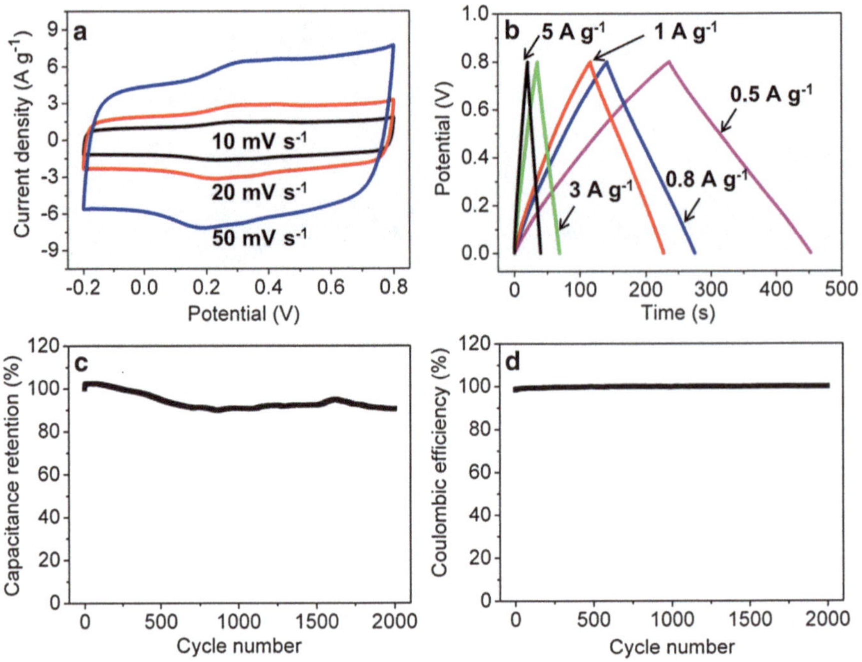

Fig. 9.11 Electrochemical properties of a supercapacitor fabric with PANI weight percentage of 50 %. **a** CV curves at different scan rates. **b** Galvanostatic charge/discharge curves at different current densities. **c** and **d** Dependence of specific capacitance and Coulombic efficiency on cycle number at 1 A g^{-1}, respectively (Reproduced from Ref. [15] by permission of John Wiley & Sons Ltd)

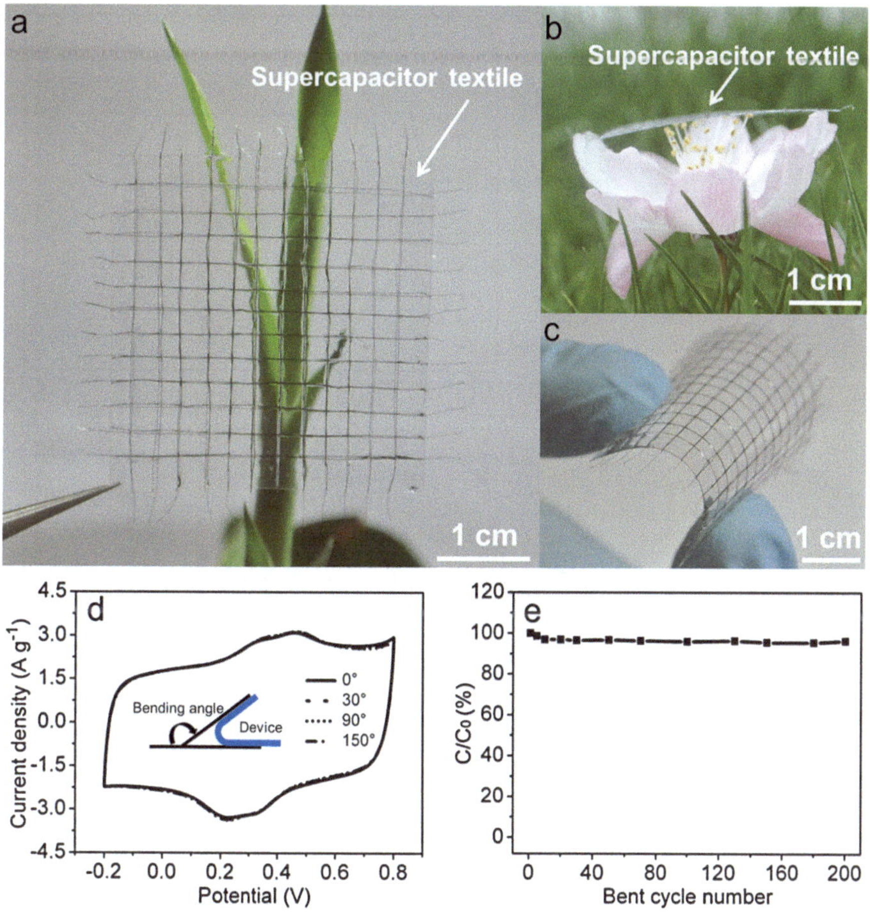

Fig. 9.12 **a–c** Photograph of a transparent supercapacitor fabric. **d** CV curves (at a scan rate of 20 mv s^{-1}) of a supercapacitor fabric being bent into different angles. **e** Dependence of specific capacitance on bent cycle number with PANI weight percentage of 50 % (bending angle of 150°). C_0 and C correspond to the specific capacitances before and after bending, respectively (Reproduced from Ref. [15] by permission of John Wiley & Sons Ltd)

Due to the grid structure and excellent flexibility of the CNT fiber fabric, the supercapacitor fabric has many merits including transparence, light weight, and flexibility (Fig. 9.12a–c). As shown in Fig. 9.12d, e, the supercapacitor fabric is sustainable to deformation without sacrificing in performance. 96.4 % of specific capacitance survived after bending 200 cycles [15].

The potential window of an individual supercapacitor is limited by water-contained electrolyte. Thus, connecting supercapacitors in series or in parallel can achieve a higher voltage and capacity. Figure 9.13a compared the galvanostatic charge/discharge curves of single device and three devices connected in series at same charge/discharge current. The potential window is extended from 0.8 V for a

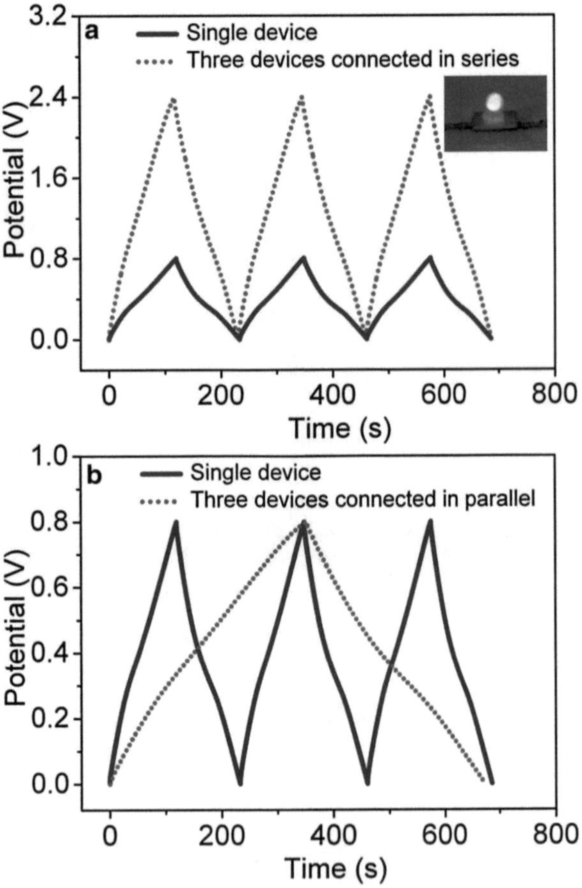

Fig. 9.13 Galvanostatic charge/discharge curves of three supercapacitors connected in series (**a**) and parallel (**b**). The inserted photograph at **a** showed a red LED powered by a supercapacitor fabric (Reproduced from Ref. [15] by permission of John Wiley & Sons Ltd)

single device to 2.4 V for the three devices connected in series. As shown in Fig. 9.13b, as the three devices connected in parallel, the potential window of the combined device maintains the same as the single device, while the charge/discharge time is tripled that of the single devices.

9.3 Perspective

Energy textiles, especially the photovoltaic textiles and supercapacitor textiles, represent advantages of high flexibility, weaving quality, and easy accessibility, which is suitable for wearable applications. But challenge still remains. The fabric electrode should be improved. Although the CNT fiber fabrics are effective for the solar cell and supercapacitor, it is difficult for large-scale application. Energy conversion and storage are two independent units; how to connect them effectively is a key step for real applications.

References

1. Aricò AS, Bruce P, Scrosati B, Tarascon J-M, Van Schalkwijk W (2005) Nanostructured materials for advanced energy conversion and storage devices. Nat Mater 4(5):366–377
2. Wang ZL (2012) Progress in piezotronics and piezo-phototronics. Adv Mater 24(34):4632–4646
3. Pan S, Yang Z, Li H, Qiu L, Sun H, Peng H (2013) Efficient dye-sensitized photovoltaic wires based on an organic redox electrolyte. J Am Chem Soc 135(29):10622–10625
4. Chen T, Qiu L, Yang Z, Peng H (2013) Novel solar cells in a wire format. Chem Soc Rev 42(12):5031–5041
5. Pan S, Zhang Z, Weng W, Lin H, Yang Z, Peng H (2014) Miniature wire-shaped solar cells, electrochemical capacitors and lithium-ion batteries. Mater Today 17(6):276–284
6. Pan S, Yang Z, Chen P, Fang X, Guan G, Zhang Z, Deng J, Peng H (2014) Carbon nanostructured fibers as counter electrodes in wire-shaped dye-sensitized solar cells. J Phys Chem C 118(30):16419–16425
7. Pan S, Yang Z, Chen P, Deng J, Li H, Peng H (2014) Wearable solar cells by stacking textile electrodes. Angew Chem Int Ed 53(24):6110–6114
8. Ren J, Zhang Y, Bai W, Chen X, Zhang Z, Fang X, Weng W, Wang Y, Peng H (2014) Elastic and wearable wire-shaped lithium-ion battery with high electrochemical performance. Angew Chem Int Ed 53(30):7864–7869
9. Zhang Z, Yang Z, Wu Z, Guan G, Pan S, Zhang Y, Li H, Deng J, Sun B, Peng H (2014) Weaving efficient polymer solar cell wires into flexible power textiles. Adv Energy Mater. doi:10.1002/aenm.201301750
10. Sun H, You X, Deng J, Chen X, Yang Z, Chen P, Fang X, Peng H (2014) A twisted wire-shaped dual-function energy device for photoelectric conversion and electrochemical storage. Angew Chem Int Ed 53(26):6664–6668
11. Chen X, Lin H, Deng J, Zhang Y, Sun X, Chen P, Fang X, Zhang Z, Guan G, Peng H (2014) Electrochromic fiber-shaped supercapacitors. Adv Mater. doi:10.1002/adma.201403243
12. Zhang Z, Li X, Guan G, Pan S, Zhu Z, Ren D, Peng H (2014) A lightweight polymer solar cell textile that functions when illuminated from either side. Angew Chem Int Ed. doi:10.1002/anie.201407688
13. Hu L, Pasta M, Mantia FL, Cui L, Jeong S, Deshazer HD, Choi JW, Han SM, Cui Y (2010) Stretchable, porous, and conductive energy textiles. Nano Lett 10(2):708–714
14. Bao L, Li X (2012) Towards textile energy storage from cotton T-shirts. Adv Mater 24(24):3246–3252
15. Pan S, Lin H, Deng J, Chen P, Chen X, Yang Z, Peng H (2014) Novel wearable energy devices based on aligned carbon nanotube fiber textiles. Adv Energy Mater. doi:10.1002/aenm.201401438

Chapter 10
Summary and Outlook

In this book, we presented a class of devices in a wire format including fiber-shaped photovoltaic devices, fiber-shaped energy storage devices, and fiber-shaped integrated devices. We also introduced the energy textiles as a demonstration to the application of fiber-shaped devices. As wearable devices become a buzzword in our daily language, energy devices pertaining to portable, flexible, and wearable electronics have been regarded as the state-of-the-art technologies. As the traditional planar devices lose their lure, an upsurge in interest has painted an encouraging picture of the fiber-shaped devices. Technically, fiber-shaped devices are derived from their planar counterparts and share the similar mechanisms, but are differentiated by the fiber electrodes. The fiber electrode plays a pivotal role in fiber-shaped devices. Tracing the development of fiber-shaped devices, it is not hard to understand that the revolution in fiber electrode always triggers the update of performance. In this book, we particularly highlight many fiber-shaped devices based on a novel fiber comprising aligned carbon nanotubes (CNTs). The CNT fiber is endowed with many merits, making it overweigh other competitors. Edging ahead of its rivals, the CNT fiber-based electrode still dominates in the field of the high-performance fiber-shaped devices, and its derivative electrode still provides a strong impetus to the innovation of devices. Here, in the last chapter, we will end our discussion in fiber-shaped devices by praising advantages, envisaging applications, and then presenting challenges.

10.1 Advantages

First materialized in 2001, fiber-shaped devices have attracted more and more attentions due to some unique advantages which live up to the requirements for modern electronics. The advantages of fiber-shaped devices are listed as follow:

H. Peng, *Fiber-Shaped Energy Harvesting and Storage Devices*, Nanostructure Science and Technology, DOI 10.1007/978-3-662-45744-3_10

10.1.1 Flexibility

Based on flexible fiber electrodes, fiber-shaped devices generally have high flexibility attributing to the one-dimensional configuration, which is favorable to wearable and portable devices. The flexibility makes the fiber-shaped device sustainable to deformations without obvious degradation in performance.

10.1.2 Miniaturization

It is widely recognized that miniaturization represents the mainstream of modern electronics. Smaller and lighter electronic devices also raise higher requirements to the weight, volume, and performance of the related energy harvesting and storage parts. To this end, conventional planar energy harvesting and storage devices can hardly meet the requirements, due to the difficulty in fabrication technique. The diameters of the fiber electrodes generally range from 0.01 to 1 mm, and the overall diameter of the resulting device can be smaller than 0.1 mm by a precise control in fabrication technique. In terms of device length, it can be well controlled on millimeter level without sacrificing the performance.

10.1.3 Weavability

Another important advantage of fiber-shaped devices lies on the weavability into fabric using traditional textile technology. Fiber-shaped devices can be woven into textiles to realize scale-up fabrication to harvest or store more energy. In addition, by controlling the connection of different fiber-shaped devices, i.e., in series or in parallel, the overall output voltage and energy can be largely increased, which is efficient for applications with high output voltage and power requirements.

10.1.4 Wearability

Compared with their planar counterparts, fiber-shaped devices are more suitable in the field of wearable electronics. Wearable devices should be highly flexible to accommodate the frequent strain during body movement; they have to be lightweight to ease the burden; they should be readily woven into clothes with stable configuration and performance. The good news is, some progress have been achieved to successfully realize wearable devices using fiber-shaped devices.

10.1.5 Others

Apart from the above advantages, certain fiber-shaped devices may share unique advantages based on the working mechanism and the one-dimensional configuration. For example, the performance of fiber-shaped solar cells is insusceptible to the direction of incident light, while planar solar cells can only accept incident light from the photoanode side or the efficiency will be largely decreased. This advantage has made fiber-shaped solar cells very promising as three-dimensional light harvesters to accept lights from different directions, especially for indoor applications. The fiber electrode in fiber-shaped lithium ion batteries is exempt from binder and conducting agent which is beneficial for energy density.

10.2 Applications

Imagined long before, materialized decade ago, the fiber-shaped devices have a promising future in applications:

10.2.1 Portable Devices

Based on their lightweight and flexibility, fiber-shaped solar cells can be used as flexible and lightweight power source to directly power or charge portable devices such as cameras, mobile phones, and laptops, etc. Fiber-shaped electrochemical capacitor and lithium ion battery can provide the pre-charged energy to power or charge portable devices. Especially with the prosperous development of flexible devices, such as flexible Galaxy Round by Samsung and Google Glass by Google, highly flexible devices for energy harvesting and storage are even more demanded, which provides plenty opportunities for fiber-shaped devices.

10.2.2 Miniature Devices

Nowadays, electronic devices are becoming smaller and lighter, which not only raises the requirements for the devices themselves but also for the batteries and power sources. For some specific miniature devices such as micromotors (diameter of 2–4 mm), the driving voltage and current is quite low (3 V, 10 mA) and the device weight is only ~0.2 g. To fully use the advantages of micromotor, the volume and weight of the power source should be taken into consideration. Fiber-shaped devices can be promising candidates as the power source to drive the microdevice, as well as other miniature devices in modern electronics.

10.2.3 Military Applications

Fiber-shaped devices are also appealing to military applications, in which the devices should be lightweight, flexible, integrated, tough, and efficient. Especially in some severe conditions, soldiers need highly integrated devices with multifunctions. To this end, fiber-shaped devices can provide good solutions based on their light weight, flexibility, weavability, and easy integration. Fiber-shaped integrated devices can be used as self-powered devices to harvest and store energy as long as there is incident light. When needed, the stored energy can be applied to drive or charge certain devices. Military tent woven by the fiber-shaped integrated devices can harvest light at daytime and power lights, computers, or communication devices at night. Armors with fiber-shaped devices can harvest and in situ store energy to supply energy to portable devices. In the future, more functions may be integrated into one fiber-shaped device, for example, the sensor function to monitor the circumstance, the memory function to record information, etc., to promise multifunction products in this field.

10.2.4 Wearable Applications

One of the most attracting applications for fiber-shaped devices lies on the potential for wearable applications. The advantages of flexibility and weavability have promised the possibility of the fabrication of energy textiles by weaving fiber-shaped devices with clothes or themselves together. Besides integrating fiber-shaped devices into clothing, other wearable devices including electronic glass and watch can also use fiber-shaped devices to realize energy harvesting and storage.

10.3 Challenges and Directions

Aware of the fact that the future of the fiber-shaped devices is still clouded by some formidable challenges, it is too early to acclaim the success.

10.3.1 Fiber Electrode

As discussed in Chap. 2, fiber electrodes play a critical role in determining the performance of fiber-shaped devices. For all the fiber-shaped devices, fiber electrodes are used to transfer electrons and should be highly conductive to reduce the internal resistance of the device. Apart from this, different types of devices ask the fiber electrode for specific features, e.g., catalytic activity in dye-sensitized solar cell, large specific surface area in electrochemical capacitor, and conductivity in lithium ion battery. However, the present existed fiber electrodes can hardly meet the requirements, which limits the further improvements of device performance. Metal wire electrodes show relatively high conductivity (on the level of 10^5 S cm^{-1}), but the specific surface area is quite small, as well as an inferior loading capability. Carbon-based fiber electrodes are superior in specific surface area and loading capability, while the conductivity is dwarfed (on the level of 10^2–10^3 S cm^{-1}). To this end, developing novel fiber electrodes that combine the advantages of different materials with high performance is very meaningful and will be one of the development directions in this field.

10.3.2 Capacity of Energy Harvesting and Storage

Although the performances of fiber-shaped devices have been largely improved, there is still much space between the fiber-shaped and planar devices. For instance, the highest energy conversion efficiency of fiber-shaped polymer solar cell is 3.81 %, which is much lower than 11.5 % of the planar counterpart. For fiber-shaped electrochemical capacitors, the highest specific mass capacitance is ~300 F g^{-1}, while the planar counterpart achieves ~3,000 F g^{-1}, 10 times larger than fiber-shaped device. To further improve the performances is very important for the practical applications of fiber-shaped devices and is obviously another important aspect for future development.

10.3.3 Stability

Considering the practical applications of fiber-shaped devices, stability is a significant factor that directly determines the lifetime of a device. The stability is largely determined by the fabrication technique, and up to date, a challenge lies on the conscious and standard control during fabrication. For example, TiO_2-nanoparticles-based photoanode has long suffered from the TiO_2 morphology, which largely decreases the stability of the resulting fiber-shaped devices. In addition, a uniform layer of gel electrolyte is critical for fiber-shaped electrochemical capacitor. If the layer is not uniform or too thick, the two electrodes will contact with each other during use, and the whole fiber-shaped device cannot be operated normally. Such technique problems are very common in the fabrication process for fiber-shaped devices and can be considered to be a direction for future development.

10.3.4 Safety

Safety is another challenge that fiber-shaped devices face. Some of the electrolytes used in fiber-shaped devices are corrosive, and some are even poisonous, which calls for strict sealing process during fabrication. In addition, the working voltage of in-series fiber-shaped devices can be higher than the safe voltage of human body, and a safe connecting and extracting process of the electrodes is very important and challenging, which is merely investigated and considered. For the future development, more efforts should be paid to increase the security of fiber-shaped devices.

10.3.5 Scale-Up Fabrication

The scale-up fabrication can be regarded as the biggest challenge for fiber-shaped devices, which has to be performed to realize the practical applications. Just as other chapters mentioned, a single fiber-shaped device can only harvest or store very small amount of energy with small voltage and current. The scale-up fabrication can efficiently increase the voltage, current, and overall energy for practical applications. However, the scale-up fabrication involves various problems, including some of the challenges mentioned above, i.e., improving the properties of fiber electrodes, optimizing the fabrication techniques, and increasing the stability and security of fiber-shaped devices. In addition, one of the most promising approaches for scale-up fabrication of fiber-shaped devices lies on the weavability of fiber-shaped devices. As a result, to weave fiber-shaped devices into textile to scale up the fabrication, especially using the mature textile technology, is very important while challenging for the state of art of this field. To this end, one of the most important directions for the future development of fiber-shaped devices is the scale-up fabrication of fiber-shaped devices.

To summarize, since the first realization of fiber-shaped device in 2001, various advancements in electrode material, device structure, device performance, and scale-up fabrication have been achieved, inspired by the advantages of flexibility, miniaturization, weavability, wearability, etc. The development of the fiber-shaped device field is still suffering from some challenges involving the fiber electrode, device performance, stability, safety, and scale-up fabrication. These challenges can be also considered to be the future directions for fiber-shaped devices, which can be further improved and optimized. We believe that with continuous efforts, fiber-shaped devices can appear in our daily lives and make our life more convenient in the near future.

MIX
Papier aus verantwortungsvollen Quellen
Paper from responsible sources
FSC® C105338
FSC
www.fsc.org

If you have any concerns about our products,
you can contact us on
ProductSafety@springernature.com

In case Publisher is established outside the EU,
the EU authorized representative is:
Springer Nature Customer Service Center GmbH
Europaplatz 3, 69115 Heidelberg, Germany

Printed by Libri Plureos GmbH
in Hamburg, Germany